Earth Surface
Sediment Transport

IAN STATHAM

CLARENDON PRESS　　OXFORD

Oxford University Press, Walton Street, Oxford OX2 6DP

OXFORD LONDON GLASGOW NEW YORK
TORONTO MELBOURNE WELLINGTON CAPE TOWN
NAIROBI DAR ES SALAAM KUALA LUMPUR
SINGAPORE JAKARTA HONG KONG TOKYO
DELHI BOMBAY CALCUTTA MADRAS KARACHI

First published 1977
Reprinted with corrections 1979

British Library Cataloguing in Publication Data

Statham, Ian
  Earth surface sediment transport.
  1. Sediment transport 2. Geophysics
  I. Title
  551.3′03   QE571   77–30295

  ISBN 0–19–874076–X
  ISBN 0–19 874077–8 Pbk

Filmset in 'Monophoto' Times 10 on 12 pt and
printed in Great Britain by
Richard Clay (The Chaucer Press), Ltd,
Bungay, Suffolk

Contemporary Problems in Geography

The general editor of *Contemporary Problems in Geography* is Dr. William Birch, who is Director of the Bristol Polytechnic. He was formerly on the staff of the University of Bristol and the Graduate School of Geography at Clark University in the U.S.A. and he has been Chairman of the Department of Geography in the University of Toronto and Professor of Geography at the University of Leeds. He was President of the Institute of British Geographers for 1976.

Alan Wilson is Professor of Regional and Urban Geography at the University of Leeds. After reading mathematics at Cambridge he has served as Scientific Officer at the National Institute of Research in Nuclear Science, Research Officer at the Institute of Economics and Statistics, University of Oxford, Mathematical Adviser at the Ministry of Transport and Assistant Director of the Centre for Environmental Studies. His publications include *Entropy in Urban and Regional Modelling, Papers in Urban and Regional Analysis* and *Urban and Regional Models in Geography and Planning*.

Michael Kirkby is Professor of Physical Geography at the University of Leeds. He has done research at Cambridge and at the Johns Hopkins University and previously taught at the University of Bristol. In 1972 he published *Hillslope Form and Process* with M. A. Carson.

To Marg and Simon,
and to my parents

Editorial Preface

The progress of research in physical geography has been accompanied, over the last twenty years, by a progressive shift in teaching content at all levels. The changes have filtered down from final year to first year courses in higher education, and increasingly to sixth form work. Suitable textbooks have been rather slow to follow, and it is only in the last five years that the trickle of advanced-level texts has grown to a relative flood. First and second year undergraduate courses are still poorly served, even though the need for course texts is much more important at this level.

Contemporary Problems in Geography is a series which is intended to fill this gap for geography as a whole; and this is one of three volumes which are being published this year on systems of interest to physical geographers. Andrew Goudie's book *Environmental Change* is an introduction to the framework of recent earth-history which is a crucial context for all studies of current process or landform. Ian Statham's book *Earth Surface Sediment Transport* examines the processes which move sediment on hillslopes and in rivers, and so form the landscape. Steve Trudgill has written *Soil and Vegetation Systems*, a book on the processes by which the plants and soil interact to provide the conditions for plant growth, and determine the course of soil evolution.

All three books are intended for first and second year undergraduates who have done little or no specialized physical geography before, and who need a broad understanding which will fit in with other aspects of geography or environmental science.

March 77

Preface

In writing a short series-text on as wide a subject as transport processes I was faced with considerable difficulty about what approach to take. Accordingly I feel I should preface the text with an explanation of how I arrived at the final contents. My aim was to introduce the basic mechanics and chemistry of sediment and solute transport at the earth's surface. I have confined myself closely to the processes involved and refrained from straying into their implications, for example to slope or drainage basin form. In the limited space available it was impossible to discuss all processes in great detail and so I attempted to group together processes or regimes of processes according to similarity of operation. They fall into these major groups; mass movements (rapid and slow); fluid sediment transport processes and solute transport. Within these major groups basic mechanisms have been discussed, though differences between specific environments, for example between different types of fluid, have only been covered superficially. It is important to stress that the text is not intended to provide exhaustive coverage of all transport environments, but to show how basic principles can be used to enhance understanding of processes.

I.S.

Oxhey, 1977.

Acknowledgments

I would like to thank my colleagues Mike Kirkby, Derek Blundell and Dave Huddart, for reading the manuscript and offering helpful criticism and encouragement. Thanks are also due to Janet Sullivan, who typed the original manuscript.

Contents

1 The Mechanics and Dynamics of Material Transport

1.1. Introduction

A river in flood is a vivid manifestation of sediment transport. Its turbid waters are carrying a burden of material derived from the land surface, a burden which amounts to millions of tons every year transported to the sea along the world's rivers. And in addition to the visually apparent sediment, there is an invisible load of material carried in solution which is almost as large. What one observes in the flooding river is the result of a wide range of material derivation and movement processes, many of which are in themselves slow or imperceptible to the human eye. These processes of material transport form the subject-matter of this book. The approach taken is systematic, in that processes which have fundamental similarities of operation are grouped together. A distinction is made between solutional transport (sections 3.2, 6.4) and sediment transport, since one is essentially chemical and the other mechanical. Another division thought to be significant is that between processes in which a lot of sediment moves *en masse* (*mass movement*, chapters 4, 5) and those in which sediment moves as individual grains more or less dispersed in a fluid (*fluid transport*, chapter 6). The distinction is thought to be justifiable because in the first case net movement is largely related to a component of the sediment's own weight whereas, in the latter, motion of an external fluid agency is the major control. The distinction is by no means entirely clear-cut; for example, it will be seen that certain slow mass movements resemble flows in detail. Also, sediment becomes increasingly fluid in behaviour through the continuous transition from dry material to muddy water. In nature these grey areas are inevitable but they should not be used as an argument against classifying processes in a way which simplifies understanding. At the same time however, one should be aware of the limitations of a classification which draws clear boundaries between what are essentially members of a continuum.

It would be fair to ask the question at this stage: 'Why study transport processes at all?' On an immediate and practical level they are relevant to land-use. Flowing water ruins much agricultural land either by eroding the soil from it or by covering it with unwanted sediment derived from elsewhere. Landslides and other soil movements damage buildings and roads by disturbing their foundations or by physically burying them. These are just two examples of the countless ways in which natural transport processes impinge on man's activities. Theoretical and empirical studies of processes have, through a clearer understanding of their operation, enabled many to be controlled and to be accommodated in designs. It is not surprising that an overwhelming contribution has been made by engineers in this respect. They are called upon to give immediate answers and long-term solutions to problems as they occur, and consequently have generally been confined to issues which markedly influence man. They have been especially active in the fields of landslides and sediment movement in water, where they have contributed the major proportion of theoretical knowledge, but also elsewhere.

On a different level sediment movement is central to the academic discipline of geomorphology, which is concerned with explaining landscape. It is outside the scope of this book to enter into a discussion of how this is achieved by geomorphologists, but it is perhaps worth mentioning that methods of approach have changed over recent years. Geomorphic work of the early twentieth century, characterized by that of W. M. Davis, treated landscape in a cyclic, evolutionary manner. It was assumed that an initial surface, uplifted by tectonism, was steadily eroded until a flat surface at a lower elevation was approached. The cycle could be restarted by another period of uplift. The strength and weakness of this approach lie in its generalization. Landscape was explicable historically, as a relative stage of development in a definable sequence. Davis's generalizations on landscape development have not yet been improved on substantially, but their weakness is that it is not possible to particularize from them. To put it another way, one cannot answer questions such as '*How* did this landscape arrive at this stage?' by the Davisian approach. There are other objections to his approach, but this one is probably the most important. One can only answer the 'how?' question by looking at the processes of material transport responsible for moulding landscape. Davisian geomorphology regards the precise nature of material transport as irrelevant to landscape evolution. Currently, geomorphologists are actively studying the action of transport processes, in pursuit of the academic goal of explaining landscapes. For the most part explanation is restricted to single processes in small areas, but some attempts at generalizations on the scale of Davis's have already been made on the basis of process studies.

One should not neglect the role of transport processes in sedimentary geology and sedimentology. Here, geologists are attempting to reconstruct environments of deposition of past sediments. Following the principle of uniformitarianism, present-day processes are taken to be characteristic of those of the past and are used to answer such questions as 'Which direction did the flow come from which deposited this sediment?' and 'How fast was it?'. Consequently there has also been a large contribution to the understanding of transport processes from geologists.

1.2. Material transport processes defined
During the operation of the geological cycle, rocks and sediments produced from magmas within the earth's crust or in depositional environments at the

earth's surface are exposed to the range of subaerial processes known collectively as weathering. Weathering processes operate in many different
ways to change the state of materials exposed at the surface of the earth, a
change in state (in a subaerial environment) being brought about by the
instability of most rocks derived from or lithified within the crust. Generally
speaking, the effect of weathering is (chemically or mechanically) to reduce
the strength of rocks and sediments and to render them more easily movable
from one location to another. The processes of weathering and material
movement or transport collectively constitute a major part of geomorphology. The sphere of interest of the geomorphologist represents a small
but extremely important part of the general geological cycle; important because it is the part of the cycle which interacts most completely with other
major flow systems, for example the biosphere and the hydrological cycle
(Fig. 1.1), and because the effects of material transport often directly influence man's activities, as discussed in 1.1. Geomorphology is then con-

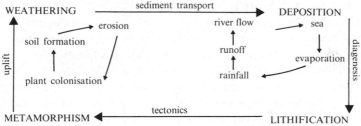

FIG. 1.1 Interaction of the geologic, hydrologic and plant-colonization cycles.

cerned with the production of material by weathering processes, the transfer
of that material down an energy gradient to a new location, and the effects
these processes have on the earth's surface as a whole. Thus a system is
under consideration in which flow of solid (*sediment*) and dissolved (*solutes*)
weathered bedrock is the most important interaction between one part of
the system and another.

The effects of weathering and transfer processes are extremely numerous.
Their mutual interactions are, for instance, important in the development of
soil materials in which plants may grow. Similarly, the spatial distribution of
sedimentary materials, produced by weathering and moved by transfer processes, determines surface landform and the sequential development of landforms through time. It should be emphasized as well that the effects of
weathering and transfer may in turn act to modify the initial processes by
feedback, a reaction in which a result influences the cause of an original
action. So a soil is not a simple product of weathering processes in which
plant growth may occur; more realistically the processes of weathering and
plant colonization are inextricably bound together in a complex system whereby weathering stimulates biotic activity and vice versa. Indeed, at many
weathering interfaces the influence of the biota is primary in importance.

For a material to be rendered transportable it must first undergo an increase in chemical or mechanical mobility or, in other words, it must be changed from a stable to an unstable substance with respect to the forces acting upon it. Weathering may reduce grainsize of a sedimentary rock, for example, and thus diminish the over-all force required to move individual sedimentary particles. Alternatively, a chemical constituent of a rock may be altered from an insoluble to a soluble form, facilitating transport in aqueous solution. These examples are only at the final level of a sequence of complex weathering interactions in the sediment and solute transport system, culminating in material movement.

Variables in any system may be classified as either system or environmental variables according to their role. This division is not constant, since a variable may change its status according to the time-scale over which the system is considered. In Table 1.1 the major variables within a general sediment and solute transport system are listed according to status over a time period necessary for one particular transport event to take place. Of course, this period is highly variable, but it can generally be regarded as very short relative to the lifespan of the system itself. Variables specifically related to material movement, such as transport process, sediment quality and weathering process, are *system* variables, whereas climatic and geological factors are *environmental*. Environmental factors affect the system in that they set the scene; they do not vary throughout the timespan under consideration. The most important lines of interaction between the factors shown in Table 1.1 are illustrated diagrammatically in Fig. 1.2 and the most significant feature of this design is the strong cause-effect-feedback loop

Table 1.1

Status of variables in a general sediment system

System Variables	Environmental Variables
Weathering Process	Climatic Characteristics
Transport Process	Geology
Sediment Discharge	Topography and Slope
Solute Concentration	Vegetation
Sediment Quality	Tectonic Activity
Solute Quality	
Sediment Storage	
Fluid Discharge	
Daily Weather	

which exists between weathering, transport process, and sediment and solute qualities. The net effect of this complex set of interactions is the movement of material which is manifest as the discharge of sediment and solutes out of the system through time.

An important variable in a sediment transport system is sediment storage, from the point of view of quantity and spatial disposition of the loose, weathered material in the system. Sediment storage not only indicates the way in which a particular process operates; it is also the way in which the balance of equilibrium of a process-form relationship is maintained. A process-form relationship exists when a specific sediment transport process

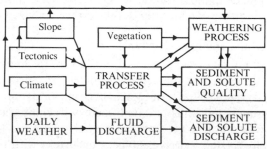

FIG. 1.2 Important interactions in a transport system.

may be demonstrably associated with a characteristic landform, and the characteristic form is often maintained by storage of sediment in the system at any given time. Sediment stored in the system is present on the land surface as superficial deposits, constituting the vast majority of the earth's surface exposed to view. Storage of sediment signifies that events in different parts of the system operate on different time-scales, causing a certain amount of intermittence in sediment movement. This intermittence may be spatial, temporal or both. The former is not difficult to explain since almost all the important environmental and system variables controlling the operation of process vary from place to place. Temporal variation in transport process rate implies that threshold or critical values in some other system variable are required to initiate the action, otherwise the process would simply operate at a slower rate. In other words a *trigger mechanism* is often required to initiate a process, and a good example is the influence that heavy rainfall and subsequent floods have on sediment movement in rivers. Indeed, many trigger mechanisms are closely related to climate, though there are other causes. Tectonic activity, for instance, is well known as an initiator of catastrophic mass movements in some areas.

Sediment transfer processes could be classified according to their spatial and temporal variability viz.:

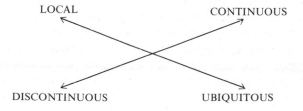

Storage of sediment is largely related to discontinuous process action. Landslides might be considered a good example of a discontinuous-localized process, whereas solutional loss from the land surface falls into the continuous-ubiquitous category. Thus, solutes are much less commonly stored on the land surface than sediment particles and, once they have been derived from the bedrock, they generally continue to move. Naturally there are many exceptions to this; redeposition of dissolved material occurs in numerous environments, but they tend to be rather localized in importance.

1.3. Scientific method and material transport

Harvey defines scientific explanation as a satisfactory (or logical) answer to a 'how' or 'why' question, or the reduction of an unexpected result to an expected one (Harvey, 1969). Open-system concepts of dynamic equilibria in geomorphic systems have necessitated increased emphasis on sediment transport processes, because the state of the system elements (e.g. landforms) are fundamentally related to process-type. Concurrent with this trend is the need to answer 'how' or 'why' questions concerning material transport, and so scientific ways of investigating these queries must be applied. Great store is placed by reason in scientific explanation, that is in the principle that one point should follow clearly from the previous one. One should not, however, forget the importance of subjective decisions and assumptions in scientific investigation. The objectives of a study, or indeed a whole discipline, are entirely based on the subjective judgement and prejudices of its operators. Assumptions are often also made to simplify a problem or to make it easier to apply known principles, and these are frequently impossible to defend in a strict logical fashion. Thus, what a scientist studies is based largely upon what he interprets as being important and upon what takes his interest. The way in which he approaches his study should, however, be logical, objective and reproducible.

Scientific studies may follow two major thought procedures in tackling problems, namely inductive and deductive reasoning. Deduction is the process whereby hypotheses and premises are proposed in such a way that subsequent hypotheses follow logically from them. This approach to problem-solving has the advantage that all which is deduced from the initial hypothesis must be true, provided of course the initial hypothesis is not in question. Its disadvantage lies in the weakness that it is never possible to prove beyond doubt the certainty of the initial premises and that the reasoning cannot be extended to phenomena not covered by the original hypothesis. In other words, one cannot use deduction to predict the outcome of an event outside the deductive framework. Induction, on the other hand, involves the inference of general truths from particular instances or events and is the basic method of constructing *models* of system behaviour. Hence, induction requires the use of probability, in which the results of a series of observations or experiments are used to predict the outcome of other events. Induction is a very valuable tool because it allows generalizations to be crystallized from complex situations, but its disadvantage lies in that one can never be certain that one's predictions concerning events will be correct. In practice, most scientific studies use deduction and induction within their frameworks. As an example, a geomorphic problem might be tackled in the following way. After the phenomenon to be studied has been identified and one's objective formally stated, a series of experiments or observations is carried out in order to obtain data on the subject in question. The data are then analysed and used to devise a model concerning the nature of the phenomenon. These are in turn used to predict into unknown areas and the model is accepted or rejected on the strength of its predictive performance. If it is rejected, one needs to examine the initial objectives, experimental techniques and conclusions from the data before designing another model.

It is emphasized that this is only one way in which phenomena may be examined in a scientific context. The above approach is empirical; that is to say, it is based upon the collection of data and upon models and hypotheses erected on the strength of experimental evidence. Hypotheses are deduced from results and are used further to construct a model of the phenomenon by an inductive reasoning process. Faults in the model may be revealed upon further experimental work, and these may stem from weaknesses in the initial experiments or from misinterpretation of evidence in the hypothesis stage. Frequently, the formal stating of an objective is replaced by an initial hypothesis, often based upon already known principles. In transport process-studies, for instance, it is a common procedure to explain process-form or process-rate relationships theoretically, using established theory from classical mechanics and dynamics. Testing of the initial *theoretical model* may of course, follow a similar pattern to that illustrated previously.

Because of the current widely held belief that processes of sediment transport are extremely important in explaining geomorphic systems, observations on process type and spatial distribution are frequently the starting points in a research programme. Experiments and observations on the processes themselves (that is, after the study-objectives have been stated) often follow two courses; viz. process-rate observations or process-mechanics observations, simply the way in which a process operates and what controls it. Rate observations are usually carried out within some sort of carefully designed spatial framework; hence, one might observe transport rates along a linear two-dimensional framework, such as a hillslope profile or a beach, or one might choose a more complex three-dimensional pattern to observe variations in rate over a wide area. The distribution of rate observations supplies evidence of form development through time since, if we know how much material is being lost or gained at a series of points in the study area, we can draw conclusions on how the surface form is changing. We can also suggest whether the present form is in equilibrium with the processes observed because, if it is changing through time, equilibrium cannot have been established. Observations on process rate do not, however, tell us very much about how or why the process operates, except that qualitatively we may find that rate is related to some other system variable such as slope angle. Process remains a 'black box', in so far as its mode of operation is an unknown. Process-rate observations are then almost entirely empirical; no application of theoretical principles is necessary. In contrast, observations and experiments on processes usually require an initial theoretical hypothesis, based upon the presumed mechanical or chemical controls, before the experimental programme can be designed. Information gathered on process mechanics and controls can also be used to predict process-form relationships, though in this case process is a 'white box' and no direct measurement of rate is called for. Indeed, a full understanding of a process and its controls could allow predictions of rate to be made for any given locality for which details of the variables controlling that process are known. Ideally, process-mechanics and process-rate studies are combined within a scientific programme, which may be broadly as follows:

(1) Theoretical hypothesis of process operation is devised, based on mechanical (or chemical) principles. Assumptions may be made about the process or the environment in which it operates in order to simplify the initial hypothesis as far as is reasonable.
(2) Experiments are designed to test whether the theory is applicable to the process. If the fit between theory and experimental results is acceptable (level of acceptance is fixed by the operator, bearing in mind what the final model will be used for), a process-form model is devised. If theory is rejected, it is necessary to reconsider the theoretical approach or the validity of the experiments.
(3) Predict equilibrium form for the system and rates of transport at points within the framework chosen.
(4) Test rate and form predictions with direct observations in the field. If the fit is not acceptable examine assumptions made in constructing the theory, and the experimental design.

Stages one to four above are idealized from the reality of geomorphic investigations into transport processes. It may not be possible, for example, to recognize the underlying mechanical or chemical principles governing the process, and the initial theoretical hypothesis would have to be replaced by a qualitative statement of process behaviour. It may often prove impossible to make even qualitative statements about the process operation and one would be forced to confine one's activities to an empirical approach, based on direct field measurements. In many cases, therefore, it may be more satisfactory to work back from the empirical study to the theory underlying the process, since the empirical evidence may give clues as to what actually takes place.

1.4. The mechanics underlying transport

All material transport, whether the material is dissolved from the action of solvents on rocks and soils or whether it consists of solid sediment particles, requires the application of a force to initiate and maintain the motion. Boulders fall from cliffs, water flows along river channels and glaciers push material before them, all due to the action of forces. Thus, in order to understand the processes of transport, one must be aware of the underlying mechanical principles which control movement. Much of the following section will be quite familiar to those who already have a knowledge of elementary mechanics, but since these fundamental principles are of prime importance to sediment transport, their inclusion is justified here for those who have not hitherto encountered them.

The mechanics of motion is divided into two convenient fields: statics and dynamics. Statics is the study of forces acting upon an object, or *body*, at rest and is thus concerned with an analysis of the conditions leading to *stability*. If a body is stable, there is no net tendency for it to move and so it will remain in its original position. The stability of a static body is definable then, entirely in terms of the forces acting upon it. Dynamics on the other hand is the study of bodies in motion and is sometimes split into two further branches: kinematics, which is the geometry of motion; and kinetics, or the analysis of forces required to promote motion. There is no fundamental difference between statics and dynamics, except that the motion is required in the dynamic case to establish the balance of forces which always exists in nature. We shall return to the balance of forces with respect to equilibrium in mechanical systems later.

Force

Motion cannot be induced in a body without the application of a force and, except in certain theoretical cases, that motion cannot be maintained without the continued exertion of some force. Equally, and less obviously at first sight, a body at rest is also subject to forces which are in balance when it is stationary. Forces are therefore responsible for promoting and preventing movement, and it is clear that they are very important in determining whether a process of material transport may take place.

A force is an action in a specified direction which tends to change the state of motion of a body and it is always balanced by an equal and opposite force called the *reaction*. If we imagine a boulder resting upon the ground, as in Fig. 1.3, it is quite apparent that it exerts a vertical force on the ground due to its own weight (W). Equally, the ground must exert a force of the same magnitude in the opposite direction on the boulder (R); if this were not the case, the boulder would sink into the ground due to its weight. The force R must exist in order to maintain the static balance of the boulder and it is important to emphasize that the reaction is not a conceptual force introduced merely to complete an equilibrium required by definition. It is a real force and its existence can be verified quite easily. For example, if you hold a weight in your fully outstretched hand it is apparent that extra tension is needed in your arm muscles to support it. Now if the weight is suddenly removed from your hand, evidence of the real nature of the reaction force is provided by the tendency for your arm to move upwards under that force. The upward movement of your arm is due to the fact that the muscles providing the reaction when the weight was present do not relax immediately after it is taken away. It is also important to note that even moving bodies have a balance of forces acting on them and we shall look at this problem later.

Forces are due to the acceleration of a body. Thus, if a body is not subject to an acceleration it cannot exert a force in any direction. At the earth's surface, most bodies are subject to the acceleration due to gravity and exert a force in the direction of gravity, or approximately vertically. The magnitude of the force exerted is generally the weight of the body in a static condition, though this must be qualified for moving bodies. For the time being, let us take the magnitude of a force exerted by a body to be equal to its weight, with the proviso that the definition of force must be examined in greater depth when moving objects are considered.

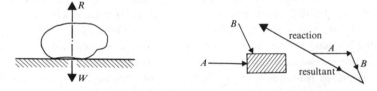

FIG. 1.3. Forces acting on a static boulder.   FIG. 1.4. The resultant of two forces.

Resultant forces

Since a force has a direction and a magnitude, we can represent it graphically as a line passing through the axis along which it acts in the body, and with a length proportional to the magnitude of the force. If more than one force acts upon a body (which is always the case) the net force due to them all is called the *resultant force*. If we return to the boulder in Fig. 1.3, it can be seen that the resultant force acting on it is zero, since the reaction is exactly equal and in an opposite direction to the weight, balancing it perfectly. There is then, no net tendency for the boulder to move. In the situation shown in Fig. 1.4, which shows the plan view of a body subjected to two forces *A* and *B*, the lengths of the arrows are drawn proportional to the magnitude of these two forces. The resultant of the forces *A* and *B* is found from the length and direction of the third side of a triangle whose other sides are equal in magnitude and direction to *A* and *B* respectively (Fig. 1.4). The triangle constructed to represent the force field is called a *triangle of forces*. Another possible graphical construction which yields the same result is to draw a parallelogram whose sides are proportional to *A* and *B*, the resultant then being represented by its long diagonal. Since the forces acting upon a body must always be in balance, the resultant force has an equal resultant reaction which acts in the opposite direction. The principle of resolution may be extended to any number of forces. Resultant forces are commonly used in solving mechanical problems in which more than one force acts on a body and problems in material-transport processes are no exception. For example, a particle of sediment entrained in the flow of a river is subject to forces pushing it vertically upwards in the flow, the force of its own weight dragging it vertically downwards, and the downstream force of the flowing water carrying it along the river channel. The magnitude and direction of the resultant of all these forces determine the net direction in which the particle will travel, and thus whether it will remain in suspension or sink to the river bed.

Resolution and force components

We have seen that two or more forces can be combined and their effect on a body expressed as a resultant force. It is also possible to reverse the process; that is to rearrange the effect of a single force to make it act in two or more directions. The forces acting in these separate directions are called *components* of the original force and they can be allowed to operate in any direction. Again, taking components of forces is extremely important in the mechanical analysis of many transport processes. The weight of a solid body is frequently the most important force tending to cause it to move, and we often want to know how much of that force acts in particular directions. If we return to our boulder resting on the ground, this time upon a slope, which is a common starting-point in the analysis of transport processes, the forces acting upon it are shown in Fig. 1.5. The weight of the boulder acts vertically in the direction of gravity but the reaction within the ground surface (R_w) prevents the boulder from moving in that direction. However, movement *down* the slope is possible because the weight of the boulder can be resolved into two forces, a force normal to the slope tending to hold the boulder in place and one parallel to the slope which tends to move it down-slope. The magnitudes of these components are found by drawing them in the required directions as the two remaining sides of a force triangle whose third side represents the boulder weight. For the case shown in Fig. 1.5, the normal and downslope components of the weight are equal to W_v and W_f respectively and their magnitudes may be calculated very simply from:

$$W_v = W \cos \alpha$$
$$W_f = W \sin \alpha$$

In just the same way as a resultant force has a resultant reaction, so a component force has a component reaction, or, in other words, the reaction to the weight of the boulder is resolved into the opposite direction to the components. Thus, W_v is balanced by the normal reaction (R_{wv}) and W_f is balanced by a parallel reaction (R_{wf}). If resistance to motion along the slope (friction) is able to supply all of the parallel reaction force there will be no downslope movement, but if the potential for resistance at the surface is exceeded the block will accelerate and the remaining reaction is made up of

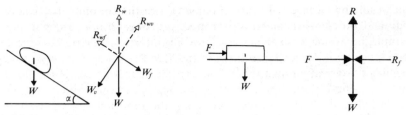

FIG. 1.5. Force components on a slope.   FIG. 1.6. Resistance to a lateral force.

an inertia force due to the block accelerating down the surface. Again, we shall discuss this in more detail under motion in bodies. Finally, resolution of forces into components is not necessarily confined to single-force systems. Components may be found for a number of forces in specified directions by first finding the total resultant force and then resolving it in the required directions.

Many transfer process systems, where combination and resolution of forces are necessary in order to analyse the mechanics of the process, can be treated as two-dimensional problems. Whether the boulder in Fig. 1.5 moves down the slope is, for instance, a question which can be answered in two dimensions working in a vertical plane parallel to the downslope direction. There are no forces operating across the slope and so that dimension is irrelevant to the solution of the problem. Very occasionally, processes of material transfer do not operate exactly along the line of steepest slope, and in these cases there is some net force component acting across the slope which must be considered. When this occurs forces must be resolved into three mutually orthogonal directions instead of two.

Moment of a force

It is often the case that a force tends not simply to push a body, but to rotate it about an axis. If such a tendency exists, the force is said to possess a moment about that axis, the magnitude (M) of which is equal to the product of the force (F) and the distance separating the line of action of the force from the axis about which it produces a rotation (d):

$$M = F \cdot d$$

For example, the force of streamflow impinging on a pebble at the bottom of a river may tend to roll it along the bed. The force due to the river flow exerts a moment on the pebble about the grain pivot point.

Forces of resistance; an introduction

Looking back to Fig. 1.5, which is a diagrammatic representation of the forces acting on a boulder on an inclined surface, it is apparent that the downslope force W_f (tending to cause the boulder to accelerate downslope) is balanced by another force known as the reaction (R_{wf}). If the block is to remain static, this reaction is a force due to the interaction between the block and the surface, which is capable of resisting motion. Resistance is obviously fundamental to transport processes since, if none existed, movement of the earth's surface material would continue under the force due to gravity until landscape became completely flat. Resistance to a force can be due to a number of mechanisms, but of overriding importance is friction. We shall have cause to look at other ways in which forces are resisted in geomorphic processes but none plays as significant a role as friction, and so it is to this we shall pay most attention for the present.

Friction exists between bodies and the surfaces over which they move, and its existence is demonstrated by the fact that most moving objects at the earth's surface eventually slow down and stop. (Newton's First Law states that every body continues in its state of rest or uniform motion in a straight line unless external forces act upon it.) Friction is not confined to solid contacts, but exists between solids and fluids and at fluid to fluid contacts. In fact, whenever differential motion occurs across an interface, we may expect there to be a certain amount of frictional resistance. Thus, a river experiences a great deal of frictional resistance at its bed and also within the water due to differential velocity of flow and turbulent eddying.

When a body slides over another body it can be shown that the frictional force is proportional to the normal reaction between them, independent of the area of contact and (to a first approximation) the velocity of sliding. Looking at the situation in Fig. 1.6, that of a block on a horizontal surface, if no lateral force is applied to the block then the frictional resistance is zero because there is no force to resist. As the force F is applied and increased the frictional force R_f increases to balance the force system. At a critical value for F the frictional resistance generated between the block and the surface will be insufficient to balance the applied force and the block will begin to accelerate. It is found that for any given surface contact:

$$\frac{F_{crit}}{R} = \text{a constant} = \mu_s$$

where μ_s is the static coefficient of sliding friction. Since the ratio is a constant, the necessary force (F_{crit}) to cause the block to move increases in proportion with R (R is the normal reaction, equal to the block weight on a horizontal surface).

A block resting on a slope, on the other hand, requires no externally applied force in order to slip. If the slope angle is great enough the downslope component of the block's weight is sufficient downslope force to cause movement. When slope angle is at the critical value, the block will begin to slide (Fig. 1.7) and

$$W \sin \alpha_{crit} = W \cos \alpha_{crit} \cdot \mu_s, \quad \text{or} \quad \mu_s = \frac{W \sin \alpha_{crit}}{W \cos \alpha_{crit}}$$

hence,

$$\mu_s = \tan \alpha_{crit} \quad \text{or} \quad \tan \varphi_{\mu s}.$$

The critical angle is called the static angle of sliding friction ($\varphi_{\mu s}$) and it is a constant, its tangent equal to the coefficient of friction, for any given surface. It should be noted that once a body has been set in motion, the force required to maintain that motion at a constant velocity may be less than the force required to initiate the movement. Hence, it may sometimes be necessary to consider a dynamic coefficient of friction (μ_k) for moving objects, whose value is somewhat less than its static counterpart.

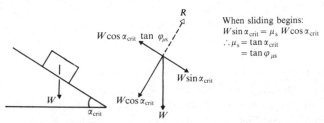

When sliding begins:
$$W \sin \alpha_{\text{crit}} = \mu_s \, W \cos \alpha_{\text{crit}}$$
$$\therefore \mu_s = \tan \alpha_{\text{crit}}$$
$$= \tan \varphi_{\mu s}$$

FIG. 1.7. Resistance to movement on an inclined surface.

Solid particles rolling over surfaces also experience a frictional resistance due to slight deformation in the surface under the weight of the body. Thus a boulder, for example, causes a slight deformation of the surface on which it lies, developing a small rut or depression into which it tends to settle. In fact, for geomorphic process purposes, rolling frictional resistance in the strict sense (where surface distortion occurs due to loading) is probably of minor importance. Almost analogous to it is, however, the rolling resistance experienced by a particle on an uneven surface and this is of major importance to the movement of sediment in some processes. Considering spherical particles resting upon an irregular surface (Fig. 1.8), when movement begins:

$$W \cdot a = F \cdot d \quad \text{or} \quad F = W \cdot a/d$$

Frictional resistance also exists within a fluid, tending to oppose flow and the passage of the fluid over and around obstructions. The way in which friction operates in a fluid is rather different from that existing between solids, since whether movement takes place or not does not rely on a critical force being applied. As long as some force is applied to the fluid, movement will occur, but the *rate* at which the movement takes place will vary with the force applied (section 2.3).

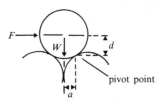

FIG. 1.8. Frictional resistance to rolling.

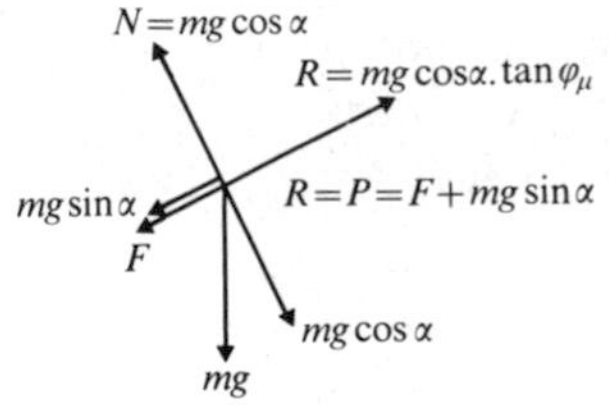

$N = mg \cos \alpha$

$R = mg \cos \alpha . \tan \varphi_\mu$

$mg \sin \alpha$

$R = P = F + mg \sin \alpha$

F

$mg \cos \alpha$

mg

FIG. 1.9. Forces acting on a boulder moving at constant velocity.

Bodies in motion

The force exerted by a static body on the ground can be considered to be equal to its weight. For moving bodies however, it is necessary to look more closely at the definition of force and, since the definition of force is related to the units of mass and weight respectively, it is necessary to examine these concepts also.

Experiments have shown that the force exerted by a body is proportional to its mass (or weight) and to the acceleration to which it is subjected. Thus:

$$F \propto m \cdot a \quad \text{or} \quad F \propto W \cdot a$$

and therefore

$$F = m \cdot a/c \quad \text{or} \quad F = W \cdot a/c$$

where F is the force exerted by the body; m is the mass of the body; W is the weight of body, a the acceleration to which the body is subjected; and c a constant of proportionality. The value of the constant c is clearly very important to the definition of a force, and this depends on whether the body is measured in units of weight or in units of mass. The *weight* of a body is (by definition) the force which it exerts at any point resulting from the acceleration due to gravity acting upon it. Thus, its weight at the earth's surface is very different from that at the surface of the moon because the moon's gravitational acceleration is much less. Mass, however, is a measure of the actual quantity of material present in the body (e.g. Volume × Density) and is not specifically a measure of force. The *mass* of the body is therefore a constant and is not dependent on its position in space.

Velocity, acceleration and the equations of motion

The rate at which a body moves in a specified direction is its velocity and is, therefore, the rate of change of distance with time (dx/dt). It is frequently necessary to know the velocity at which a transport process takes place in order to calculate the rate of removal of material.

Acceleration is the rate of change of velocity through time (dv/dt) and is an extremely important concept because of its significance in the definition of force. Mechanical problems involving moving bodies cannot be analysed unless the acceleration to which the body is subjected is known. Acceleration due to gravity is the most fundamental acceleration, and is responsible for most forces tending to move material in transport processes at the earth's surface.

There are several simple equations relating factors important in the motion of bodies and which are extremely useful when information about velocity, acceleration and position is required in the analysis of a process. If any three of the following five parameters are known: initial velocity of a body (u), final velocity (v), time of travel (t), distance travelled (s) and acceleration (a); then the unknowns may be calculated from the following formulae.

Firstly, the distance travelled by a body is simply:

$$s = \text{average velocity} \times \text{time}$$
$$= (u + v)/2t \tag{1.1}$$

provided that a is constant.

$$\text{Secondly, acceleration} = \frac{\text{change of velocity}}{\text{time}}$$

or

$$a = \frac{v - u}{t}$$

and so final velocity $= v = u + at$ (1.2)

If we substitute equation (1.2) in the form $t = \left(\dfrac{v - u}{a}\right)$ into equation (1.1)

we can derive an expression for final velocity in terms of initial velocity, acceleration and distance travelled:

$$s = \frac{1}{2}(u + v) \cdot \frac{(v - u)}{a} = \frac{(v^2 - u^2)}{2a}$$

and so $v^2 = u^2 + 2as$ (1.3)

Finally, we may obtain an expression for distance travelled in terms of time taken, acceleration and initial velocity again by substitution of equation (1.1) into (1.2) this time in the form $v = u + at$:

$$s = \frac{1}{2}(u + u + at)t$$

$$s = ut + \frac{1}{2}at^2 \tag{1.4}$$

These four simple expressions can be manipulated to solve problems involving the movement of bodies, providing they accelerate at a uniform rate.

Equilibrium in force systems

The concept of equilibrium in mechanics is a very important one and all mechanical analyses are based upon it. We have seen earlier that every force has an equal and opposite reaction and so is always balanced in nature. This principle is easily grasped in the case of a static body since the result of the force balance is expressed by the lack of motion. It is less easy to appreciate in a moving body but nevertheless balance must always exist. If a balance of forces existing on a body is taken as a definition of equilibrium, one may feel justified in concluding that the concept is of limited value since it must always be present. The conclusion would be erroneous, however, first because force systems acting on bodies would be insoluble unless a balance existed and secondly because different states of equilibrium can be recognized. These different states allow different questions to be answered concerning the movement of materials: questions such as 'What conditions are necessary for movement to take place?' and 'What conditions favour the continuation or cessation of movement?'; both critical questions in that they relate to whether a process will take place under certain circumstances and whether it will continue to operate.

A body is in *static equilibrium* when the balance of forces acting upon it requires no movement in order to restore or maintain the balance. Thus, the body remains stationary with respect to its surroundings. Clearly, much weathered bedrock and sediment is in static equilibrium at the earth's surface at any given time. It is not being moved because the forces applied to it are too small. In terms of our simple boulder on a slope then, the downslope component of its own weight would not be sufficient to overcome the potential for frictional resistance between the boulder and the slope itself. The limiting static condition is extremely important in order to assess the ability or competence of a process to transport materials. In the limiting case the forces tending to move some material are exactly equal to the forces resisting the motion and therefore the condition represents the minimum applied force that will cause some movement to take place. If the forces can be defined for the limiting static equilibrium case, one is in a position to say whether sediment movement is going to occur in a particular environment, and to place threshold limits on the operation of a particular transport process. Just as a stationary particle is in static equilibrium with respect to the forces applied to it, so a moving particle may be considered to be in *dynamic equilibrium*. There are, however, two possible conditions of motion which may occur: the body may be accelerating or it may be moving at a constant velocity. A body moving at a constant velocity may be said to be in a steady state, because no acceleration or deceleration is required to keep the forces applied to it in balance. The concept of steady state in movement is a valuable one in transport processes because the balance of forces operating on a non-accelerating moving object is that necessary just to maintain transport. Hence, if we can analyse this case for a process of sediment movement, we can say what forces need to be applied in order to ensure its continuation. The forces existing in the static and steady-state equilibrium cases therefore often tell us all we need to know about the mechanics of a sediment transport process. We have already seen that if an unbalanced force acts on a body, the balance is restored by the body accelerating in the direction of the applied force at a rate proportional to the force and inversely proportional to the mass of the body. If this is the case, a body travelling at a constant velocity must be subject to perfectly balanced forces because it shows no tendency to accelerate. Thus, the forces acting upon it are exactly the same as if the body were static and there is really no difference between the mechanics of a static body and one that is moving at constant velocity. The balance of forces acting on a boulder sliding down a slope at a uniform velocity is shown in Fig. 1.9 and it should be noted that the downslope movement force ($P = F + mg \cdot \sin \alpha$) is exactly balanced by the frictional resistance ($R = mg \cdot \cos \alpha \cdot \tan \varphi_\mu$).

An accelerating body is subject to unbalanced forces and, referring to our boulder on a slope, the reaction provided by frictional resistance is not sufficient to balance the total downslope force. The remaining reaction in the upslope direction, necessary to restore the force balance, is provided by an *inertia force*. The inertia force is a 'push' which exists between the boulder and the ground, acting in the opposite direction to the acceleration, and is equal to the product of boulder's mass and its downslope acceleration. Since no analysis is effectively possible for an accelerating situation, it will not be considered further.

Section 1.4 has been a brief review of the principles underlying mechanics and the movement of material and we shall go on to apply these principles to the movement of sediment in process events. At this stage, however, it should be noted that the application of mechanics to transport processes is rarely straightforward. It is frequently difficult to identify all the factors which influence relevant forces in a system and even if they can be isolated it is often difficult to express these factors quantitatively. A good example is the force exerted at a stream bed due to the flow of water over the bed. In order to analyse the mechanics of sediment movement at the river bed, it is necessary to assume that the water force is spread evenly over the bed (section 6.1). Observations show that this assumption is not valid; turbulence in the flow causes the force to be concentrated at some points but reduced at others, and there is no way that this effect can be accommodated easily in mechanics. Consequently, there will be frequent discrepancies between mechanical models and observed behaviour in systems which can only be reconciled in terms of probability. Mechanics can help us to understand how a process operates and whence energy is derived for the process to take place but it does not always enable us to build rigid deterministic models of how the system operates or to predict process events and rates of sediment removal.

1.5. The origin of forces in transport processes

All movements of sediment and solute-bearing water require the application of some sort of force. The force, derived from gravity, climatic effects and so on, may act on material in order to move it either directly, or indirectly through an agency. In the latter case, the force applied causes the agency to move, which in turn exerts a force on the material and tends to move it. Clearly, action through an agency is potentially less efficient; resistance to motion must be overcome in the agency as well as in the material to be moved; but it allows much greater forces to operate on sediment than might otherwise be possible.

Gravitational forces

Gravity, or the attraction that bodies possess for each other, is by far the largest source of force for process action at the earth's surface and is the only one which can act directly to move a body; that is without the aid of any force other than those derivable from the weight of the body itself. Further, gravity is ubiquitous and relatively (though not entirely) uniform in magnitude at the earth's surface.

We know that $F = m \cdot g$ (where g = acceleration due to gravity, section 1.4) and that the acceleration due to gravity has been measured experimentally, and so it is not necessary to refer to Newton's gravitational law in order to find the force acting on a body at the earth's surface. Generally, the acceleration due to gravity is 9·81 metres per second^2 (m/sec^2) though the value of this 'constant' varies with distance from the earth's centre, and with latitude since the earth is not quite a perfect sphere. Variations in the acceleration due to gravity and in the direction in which the force acts are, however, so small that they can be ignored for the purposes of transport processes. The weight of an object at the earth's surface acts vertically, towards the centre of the earth, and resistance prevents it from sinking into the ground. On a sloping surface, a *component of the body's weight* acts down the slope (section 1.4) and it is this component which provides the driving force for the vast majority of transport processes. Not only does the weight of an object provide some force to move itself on a slope, the weight of agencies such as flowing water also provides a downslope drive. Obviously, this force source is one of the most important and direct in causing material movement.

We have already seen (Fig. 1.7) that the downslope component force for a boulder on a slope is $W \cdot \sin \alpha$, where W is the weight of block and α is the slope angle. This is also true for a collection of particles, such as a soil cover on a hillslope considered as a single unit or even a depth of water flowing at constant velocity. A landslide is a good example of how the downslope component of a mass of sediment can cause movement to occur.

Fluid forces

Water flows over sloping surfaces, such as hillslopes, either as a subdivided or uniform sheet or as a concentrated channel flow in streams and rivers. Water is *fluid*, which means that if any force at all is applied to it, it will move in the direction of the applied force (section 2.3); no critical level of force is necessary. Thus, water flows downslope under the influence of its own weight, which is of course a gravitational force. Not all of the downslope force is necessary to move the water, however, and so what is left over after overcoming the various resistances to flow may be used to transport sediment in the flow and along its bed. Water also carries dissolved material, which travels at the same velocity as the water and behaves sensibly as part of the fluid itself. Thus, if the water is moving, any dissolved material in it also moves; there is no critical velocity below which dissolved material 'settles' in the flow as with sediment particles.

There are several ways in which flowing water exerts a force on sedimentary particles in order to move them. First, there is a physical 'push' due to the force exerted on a bed by the water, which can cause rolling or sliding along the bed according to their shape and roughness. Before this type of bed movement can begin, the resistance (frictional and cohesive) between the grain and its bed must be overcome. Secondly, turbulent eddying within the water causes upward and downward transfers of energy within the flow, and an upward transfer can lift grains from the bed to be transported in a suspended form. To enable this process to take place, the submerged weight of the particle must be overcome. Thirdly, aerodynamic lift may pick up grains due to a declining pressure gradient which is caused by the increase in flow velocity away from the stream bed. Whether material remains suspended in the flow after being lifted up depends on many factors. The settling rate of the particles, which increases with particle size, is one of the most important controls on whether a grain remains in suspension, but other influences such as the distribution of turbulence and the occurrence of collisions with other grains have a significant role to play.

Water flow is, then, an example of a geomorphic energy source in which gravity acts through an agency. Ultimately, the forces exerted are gravitational, but they enable material to be moved which would otherwise have been stable. Slope is, of course, an essential element in controlling the magnitude of the forces generated and water flow would cease without it.

Water pressure forces

When any object is immersed in water it experiences an upward buoyancy or *water pressure force* equal to the weight of water displaced. Thus, soil or sediment which is below the water table (the level below which pore-spaces in the ground are completely water-filled) experiences an upthrust force which tends to relieve some of the downward force of the sediment weight. On a slope, in which the water table and the direction of flow are parallel to the ground surface, the upthrust due to buoyancy is normal to the slope and is not vertical. This is a common groundwater flow condition and results in some of the slope-normal component of the sediment weight being reduced but none of the downslope component (Fig. 4.6). Hence, the effective force tending to hold the soil in place is lessened whereas that promoting its downslope movement is unaffected, and a lower slope angle is necessary to overcome frictional and cohesive resistance between grains (chapter 4). An entirely submerged slope does not show the same effect because upthrust is vertical and the two force components are reduced proportionally.

Water exists above the water table in sediments and soils, though the pore-spaces are not completely filled. Water in this zone is derived mainly from capillary rise from the saturated levels below and from downward percolation of rainwater from above. Capillary rise is a surface tension effect and the height of rise from the water table depends on the sediment grain-size. A *suction force* is caused by the rise, which is responsible for negative pore pressures existing in the grain interstices. The negative pore pressures result in the soil grains being drawn together more strongly, increasing the normal force between them and therefore their resistance to movement, familiar to us all as the force responsible for keeping sandcastles intact. Capillary cohesion, as this increase in the interparticle normal force is called, increases with decreasing grainsize since the maximum height of capillary rise is related to grainsize in a similar manner. The maximum capillary rise in a coarse sand may be only a few centimetres, and the rise will take place very quickly, whereas, in a clay, capillary rise of many metres is possible but at a very much slower rate of rise. Thus, the suction forces developed in coarse sand will be very small and probably not significant in increasing the resistance to grain movement, but in fine silts and clays the force may cause a very large increase in resistance.

Raindrops strike the ground with a force dependent upon their size and terminal velocity and in some cases this can be great enough to move particles of sediment. The force of impact is utilized in vertical consolidation of the soil, lateral pushing due to the direct impact of a drop on a particle, and finally, reflection of part of the drop as splash which also carries some material. On a horizontal surface, raindrop impact merely causes a to-and-fro movement of sediment with no net transport in any direction, but on a slope both the downslope component of the impact force and the longer travel of splashed material in the direction of slope result in a net transport of sediment (Fig. 6.5 and section 6.3). How large the raindrop-impact force is depends on the size and velocity of the drop.

Expansion forces

Many processes in sediment transport derive some of their force requirements from expansion, either in the sediment, in the solid rock itself or in enclosed pore water. Such expansions are almost entirely related to meteorologic events, such as changes in temperature and changes in soil moisture content due to rainfall and evaporation. They could therefore, be termed climatic forces. Expansion *per se* is rarely responsible for sediment transport, since its action is isotropic (acts equally in all directions), other things being equal, and very often entirely reversible. On a slope, however, gravity selectively increases expansion in the downslope direction and reduces contraction in the upslope direction, and so the process of expansion and contraction is not reversible. Here we see again the importance of gravitational forces acting through an agency.

Heating and cooling due to *solar radiation* set up stresses in coherent rock and are much quoted as a process of exfoliation or granular disintegration. Thus expansion by solar radiation provides a mechanism for overcoming the very high resistance to movement possessed by coherent rocks, and enables gravitational forces to cause transport. There has been, however, much discussion about the potency of expansion and contraction cycles to break hard rocks, largely centred about results of experiments which showed little breakdown. This issue remains in doubt but it is probable that chemical changes aid the breakdown. Solar heating can also be responsible for a progressive creeping of loose sediments and boulders over bedrock or over underlying sediment (section 5.1).

Since water expands by about 9 per cent on freezing, it may be responsible for pushing material downslope and in some cases fracturing hard rocks when freezing takes place in cracks and joints. Concentrations of ice in lens- and wedge-shaped forms are also responsible for much sedimentary disturbance, sorting of particles by size and net downslope transport of sediment, especially in areas of permanently frozen ground conditions (*permafrost*). Uniform freezing of a soil or sediment produces an expansion and heave roughly parallel with the ground surface but when thawing occurs settlement tends to be vertical (Fig. 5.5). Thus, there is a net downslope displacement of the sediment with maximum movement at the surface, which is related to both slope angle and depth of freezing (section 5.2). In some cases, migration of water takes place upwards to the freezing ground surface, resulting in the growth of ice needles which are extruded from the soil pores, rather like toothpaste from a tube. These needles, commonly 5 cm or more in length, push the surface soil grains upwards provided there is no vegetation cover to bind and protect them. As soil moisture increases and decreases, many sediments and soils expand and contract respectively, owing to the variation in the strength of the *capillary tension forces* and also to the tendency in some clay minerals to swell as they absorb water. The resulting expansion and contraction produce a similar pattern of heave and settlement to that already discussed in uniform freezing of soils, viz. a roughly parallel ground heave and vertical settlement. The mechanism is one of the most important in aiding seasonal soil creep to take place, and is considered further in chapter 5.

Less important than the sources already discussed, but still significant in some environments, are expansions due to *mineral changes*. The growth or precipitation of minerals from solution can lead to soil expansions. This is particularly prevalent in arid and semi-arid climates where evaporation at the surface causes considerable capillary rise of salt solutions, which deposit their solutes in the soil pores.

Global fluid movements

Wind is able to move sediment in a similar way to flowing water, either as a bed movement or as suspended material. Since air is a much less dense fluid, however, it is able to carry only a much smaller grainsize of sediment for the same flow velocity. Wind transport differs from water transport in other respects. Generally wind is not capable of significant erosion of a vegetation cover (there are exceptions to this) and so its effects are really confined to places where sediment is immediately exposed at the ground surface, such as deserts, high latitudes and high altitudes. Also, wind does not have to flow in channels or downslope since it is controlled by global differences in atmospheric pressure which are ultimately the result of greater heating at the equator than at the poles. Thus, wind erosion and transport are very broad in their effects and do not show much tendency to concentrate their influence along narrow flow lines.

Related to wind circulation and to the distribution of solar radiation over the globe is the movement of ocean currents. These may be surface or bottom currents and can be responsible for sediment movement on the deep ocean floor as well as on the continental shelf. Finally, tidal currents and waves along the breaker zone both produce movements of beach and sea-bed sediment. These sediment transport processes, however, fall outside the scope of this book, which is concerned only with terrestrial processes.

Biotic Forces

This section would be incomplete without a brief mention of the forces exerted by plants and animals in sediment transport. The growth of plants and plant-root systems pushes material aside, and may be in a preferred direction if a slope exists. Similarly, burrowing animals redistribute a considerable amount of sediment in some environments and again there is usually a tendency towards preferred movement downslope. These activities are discussed more fully in chapter 5, but it is worth noting here that the slope is a key factor in determining the direction net transfer takes. Once again, gravity is the underlying force of motion, acting through a biotic agency.

Conclusion

From the foregoing discussion on the derivation of force in transport processes two major points may be extracted, viz.:

(1) Slope, and therefore some downslope gravitational force component, is necessary for most movements of sediment. Gravity may act directly in this respect, and the gravitational force is derived from the weight of the material being moved, or it may act through an agency such as flowing water. There are very few exceptions to this principle; even expansion forces, for example, ultimately rely on gravity for movement to take place.

(2) Climatic and meteorologic variables are the second biggest source of energy for movement. These may cause expansions which are responsible for net movement when combined with slope or they may result in large scale global fluid movements of which winds and ocean currents are examples. These global fluid systems provide the only large-scale source of energy which does not rely on slope for net sediment movement in a given direction.

The ways in which force sources and agencies of transport interact to promote sediment movement are summarized in Fig. 1.10.

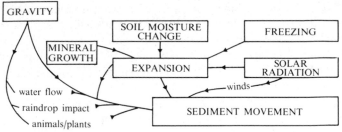

FIG. 1.10. Relationships between forces, agencies and movement of sediment.

1.6. Chemistry of material transport

Chemical Reactions

Chemical transport is the movement of dissolved rock constituents in water. In just the same way as mechanical transport requires particles of sediment to be unstable with respect to the forces acting upon them, so chemical transport relies on rock constituents being unstable in the presence of other chemicals. In the chemical sense, however, instability is a question of whether a mineral will *react* with other chemicals to form new materials. This is not essentially very different from mechanical instability since it is a response at the molecular level to applied forces. A reaction is a change in the organization of chemical constituents (*reactants*) to produce new chemicals (*products*). A reaction might take the form

$$AB + CD \rightarrow AD + CB$$

where A, B, C and D represent *ions*, which are positively or negatively charged atoms or associations of atoms. Positively charged ions are called *cations* and negatively charged *anions*. They gain their charge by the addition or subtraction of one or more negatively charged particles (*electrons*). Thus the reaction above may be written

$$A^+B^- + C^+D^- \rightleftharpoons A^+D^- + C^+B^-$$

where A^+ and C^+ are cations and D^- and B^- are anions. Notice that positively charged and negatively charged ions always combine together. The double arrow shows the reaction to be reversible, which is often the case to some extent. This reaction, between *ionized* chemicals (that is, dissociated into ions), is by no means the only way in which chemical reactions can proceed. But it is common in aqueous solutions, which is the environment in which chemical transport takes place.

The chemical equation tells us two things. It tells us the chemicals and ions which are present in the reaction (represented by standard symbols) and it tells us the proportions in which they associate with each other. This depends upon their charge, or in other words, upon the number of electrons the atom gains or loses on ionization. The charge is usually called the *valency* of an ion. For example, sodium (Na) loses one electron on ionization to gain a unit positive charge, while chlorine (Cl) gains an electron to become unit-negatively charged. They both have a valency of one and combine with each other in the ratio of 1 : 1 (by ions, not weight) to balance each other's charge:

$$Na^+ + Cl^- \rightleftharpoons Na^+Cl^-$$
sodium chloride.

On the other hand, the carbonate anion (CO_3^{--}) has a double negative charge (a valency of two) and so combines with sodium in the ratio 2 : 1:

$$2Na^+ + CO_3^{--} \rightleftharpoons Na_2^+CO_3^{--}$$
sodium carbonate

The number of electrons gained or lost by an atom on ionization need not necessarily always be the same. Iron (Fe) for example may lose two or three, and exist as Fe^{++} or Fe^{+++}, according to the conditions.

Chemical reactions take place between gases, between solids and between liquids, or between combinations of all three phases. There are really only two types of reactions which are important in chemical transport, however, viz. those between water and minerals and those between constituents dissolved in water and minerals. In the latter case, which is by far the more important, some of the dissolved constituents may be gases from the atmosphere as well as dissolved solids released from other minerals. Reactions in chemical transport are really the reactions of *chemical weathering*, which take place at rock surfaces, within soils and sediments, and at the soil : bedrock interface (*the weathering front*). Water does not necessarily play a direct part in weathering reactions but its presence appears to be essential for all of them to operate effectively. Ionic reactions take place much more rapidly in an aqueous environment than between juxtaposed solid constituents. From a practical standpoint as well, sufficient water must be present for flow to occur, otherwise no transport of dissolved materials would take place. Some weathering reactions are very simple, involving only one stage, but the majority are complex multi-stage operations with a number of intermediate products. The types of reaction important in weathering are looked at later in this section. We shall return now to the question of why reactions take place at all.

Chemical equilibrium

In the hypothetical reaction of *AB* with *CD*, how can one determine whether the reaction is possible and how far it will go if it is a reversible change? A short answer is that the reaction takes place if there are forces of attraction between *A* and *D* and between *C* and *D*, and how far the reaction goes depends on the relative strengths of all the forces of attraction between all the ions. This answer to the question of course hedges any discussion of the forces involved. It is possible to quantify the extent to which a reaction proceeds. Take a general reaction:

$$nA + mB \rightleftharpoons xC + yD$$

where n, m, x and y are integers representing the proportions in which reactants and products A, B, C and D react. It has been found that

$$[C]^x . [D]^y = K[A]^n . [B]^m \qquad (1.5)$$

where the square brackets denote concentration of the constituents and where K is a constant. In other words, the reaction proceeds to a *balance or equilibrium*, at which the forces promoting the forward reaction are equal to those driving the reverse reaction. At equilibrium the concentrations of the constituents remain constant and it is clear from equation (1.5) that if the concentration of any constituent is raised or lowered the reaction proceeds to a new equilibrium with different concentrations for all the constituents. K is called the *equilibrium constant* and varies with temperature and pressure. Consequently, it is always quoted for specified temperature and pressure conditions. The relationship in equation (1.5) is true for all reactions but here we will only be concerned with mineral solubility.

As an example of mineral solubility, if gypsum ($CaSO_4$) is added to water and stirred a reaction takes place resulting in the liberation of Ca^{++} and SO_4^{--} ions into solution:

$$CaSO_4 \rightleftharpoons Ca^{++} + SO_4^{--}.$$

The equilibrium constant for the reaction is

$$K = \frac{[Ca^{++}] . [SO_4^{--}]}{[CaSO_4]}.$$

Concentrations are expressed as gram molecular weight per litre of solution. Provided the amount of gypsum present exceeds the amount which can be dissolved, the amount dissolved reaches a constant equilibrium level, called *saturation*. Since saturation concentration is independent of $[CaSO_4]$, this may be set to unity and K becomes

$$K = [Ca^{++}] . [SO_4^{--}]$$
(the product of the Ca^{++} and SO_4^{--} concentrations).

K is often called the *solubility product* for chemicals dissolving in water and it is a direct measure of how soluble a material is. For the solution of gypsum ($K_{25 °C}$) $= 3.4 \times 10^{-5}$ and since the same number of Ca^{++} and SO_4^{--} ions must dissolve, solubility $= \sqrt{K}$. Thus, the solubility of gypsum at 25 °C in pure water is 7.8×10^{-1} gm/litre.

The solution of silica in water takes the form

$$SiO_2 + 2H_2O \rightleftharpoons H_4SiO_4,$$

silica dissolving as silicic acid, which is unionized. The solubility product here depends on crystal structure, viz.:

$K_{25°C}$ for amorphous silica $= 2 \times 10^{-3}$, solubility $= 1.9 \times 10^{-1}$ gm/litre

$K_{25°C}$ for quartz $= 1.7 \times 10^{-4}$, solubility 1.6×10^{-2} gm/litre.

It is only fair to point out that the value for quartz is an extrapolation from higher temperatures since the rate of quartz solution is negligible and crushing only produces a surface layer of amorphous silica (Berner, 1971).

The solubility product provides a way of calculating the final result of a reaction, but does not really define whether the reaction will take place or at what rate it proceeds. It is simply an empirical approach, based on careful laboratory observations. A side issue is raised here by the question of rate, concerning the definition of equilibrium. If a reaction takes place very slowly, such as the solution of quartz, how do we know when equilibrium is reached? This is a very difficult question to answer indeed, which may be solved partially by starting the reaction from both ends. If there is little tendency to proceed in either direction, it is probably a very slow reaction since the products and reactants could not be in equilibrium at both ends of a reversible reaction.

Returning to chemical reactions, since energy possessed by the atoms and molecules provides the driving force for reactions it seems reasonable to define equilibrium as

(total free energy of products) − (total free energy of reactants) = 0.

From this it is apparent that if the total free energies of the products and reactants are *not* equal, a reaction will proceed in the direction of lower free energy. Thus the tendency for a reaction to proceed depends on the *difference* in free energy (ΔF) between reactants and products. The change in free energy in reaction is due to heat input or output and to reduced capacity of the system to perform work. Free energy and solubility product are related:

$$F = -1.364 \log K.$$

The thermodynamic principles underlying this relationship and the precise definition of free energy may be followed up by reference to the texts quoted in 'Further Reading' at the end of this chapter.

It is fair to point out that although a consideration of free energy defines the tendency for a reaction to take place, it does not *necessarily* give an indication of the rate of reaction. Rate is dependent on physical parameters as well as chemical. For example, rate of diffusion of ions away from a reacting surface is important in that it controls the release of more ions. Hence, stirring increases the rate of solution of many minerals, since it causes the removal of the layer of saturated water surrounding mineral grains. Also of importance is crystal structure, demonstrated by the differing solubilities of quartz and amorphous silica.

Weathering reactions

There is only a limited range of chemical reactions causing rock weathering, owing to the relatively restricted physical environments in which weathering takes place and the range of reagents available to take place in reactions. The reactions can be conceptualized individually though it must be stressed that chemical weathering is rarely simple, often consisting of several simultaneous or sequential reactions with many intermediate products.

The simplest weathering reaction releasing soluble products is of course *solution* itself, where molecules or ions are released into water. Two examples, gypsum and silica, have already been quoted. Generally, solution is only significant on moderately to highly soluble minerals such as halite (NaCl) and gypsum. Since these are not usually present in abundance in soils and rocks and are removed quickly, solution may be regarded as a rather small contributor to chemical weathering. It must be added, however, that in some areas the highly soluble minerals form very large accumulations in rocks and soils. Most minerals, like quartz, are so sparingly soluble that direct solution causes a negligible part of their chemical weathering.

Carbonate minerals such as calcite and dolomite are relatively insoluble in pure water but owe their reactivity in natural waters to dissolved carbon dioxide gas. The *carbonation* reaction, a three-phase process between water, carbon dioxide and for example, calcium carbonate may be summarized thus:

$$CaCO_3 + (2H^+ + CO_3^{--}) \rightleftharpoons Ca^{++} + 2HCO_3^{--}$$

$$\underset{\text{carbonic acid}}{H_2CO_3} \rightleftharpoons \underset{}{H_2O} + \underset{\text{from air}}{CO_2}$$

calcite   bicarbonate ion

Calcium is released as the more soluble bicarbonate.

There are two important controls on the reaction. Firstly, increased partial pressure of carbon dioxide in the air causes more to be dissolved in water and hence pushes the carbonation reaction to the right. It is for this reason that calcium carbonate solution is higher in soils where the partial pressure of CO_2 is greater due to micro-organism respiration. Secondly, an increase in the concentration of H^+ ions, or an *acid* environment, would favour the forward reaction. The question of acidity in chemical weathering is discussed under environment, below.

Strictly speaking, *oxidation* is the process of subtracting electrons from an ion and *reduction* the reverse. In practice, many ions reach their highest valency in combination with oxygen and therefore many oxidation reactions directly involve the addition of oxygen. As far as weathering is concerned, oxidation only occurs in the presence of water and affects mainly iron and manganese minerals but also metallic sulphides. For example:

$$\underset{\substack{\text{fayalite}\\\text{(olivine)}}}{Fe_2SiO_4} + \tfrac{1}{2}O_2 + 2H_2O \rightarrow \underset{\substack{\text{iron}\\\text{oxide}}}{Fe_2O_3} + \underset{\substack{\text{soluble}\\\text{silica}}}{H_4SiO_4}$$

The iron is converted from the ferrous (Fe^{++}) state in fayalite to the ferric (Fe^{+++}) state in iron oxide, and soluble silica is released. The equation represents the result of the reaction and not the process, which probably involves intermediate stages. As would be expected, oxidation is favoured by well-aerated environments, such as freely drained soils, whereas reduction occurs in anaerobic environments, found in waterlogged conditions.

Hydrolysis is a reaction between a chemical and water to give an acidic or basic solution, unlike simple solution where neutral conditions are obtained. It is probably the most important weathering reaction because of its effect on common rock-forming minerals such as felspars. Two reactions will be sufficient to demonstrate the principle:

$$Mg_2SiO_4 + 4H_2O \rightarrow 2Mg^{++} + 4OH^- + H_4SiO_4$$

forsterite   soluble   soluble
(olivine)   magnesium   silica
                                ions

$$4KAlSi_3O_8 + 22H_2O \rightarrow 4K^+ + 4OH^- + Al_4SiO_4O_{10}(OH)_8 + 8H_4SiO_4$$

felspar   soluble   kaolinite   soluble
                          potassium   silica
                          ions

The main points to note are that metallic cations such as magnesium and potassium are released into solution, along with silica. Indeed, this reaction is probably the dominant source of soluble silica transported by rivers and a very important source of metal cations. The reaction also produces OH^- ions which tend to make the solvent more *basic* (see Weathering environments, below). Thus, the forward reaction is favoured by an acidic environment which tends to neutralize the OH^- ions released. Once again, the equations only show the result of the reaction and hide the process. The first stage is probably an irreversible substitution of H^+ ions for metal cations in the mineral lattice. In the case of felspar weathering this leaves a hydrogen felspar which later breaks down to a clay mineral, with the loss of soluble silica. Hydrogen ion substitution and the removal of silica and metal ions from soils is discussed in section 3.1.

Weathering environments

Environmental factors influence the rate and possibility of chemical reactions and can broadly be divided into physical factors and chemical factors. Physical factors include temperature, pressure and water-flow rate, and chemical factors can largely be summarized by acidity and oxidation potential.

Temperature influences the equilibrium constant for all reactions, and for solution the solubility product increases with temperature. This is a logical shift since the solution of solids in water tends to be an endothermic reaction (heat-absorbing) and so an increase in temperature will cause a shift towards solution. The reverse is true for gases, however, which cause heat emission when dissolved in water (exothermic). Thus, reactions which depend upon dissolved gases, such as carbonation, are inhibited by temperature rise. For the most part *pressure* is relatively constant throughout all weathering environments and so cannot really be considered an environmental factor in weathering. However, the *partial pressure* of some gases in the atmosphere, such as CO_2, does vary and is particularly high in soils. Thus carbonation, and other reactions which depend on CO_2 solution as a source of H^+ ions, are enhanced in soils. Finally, the rate of reaction is proportional to *water-flow rate* because increased turbulence aids the diffusion of dissolved ions away from mineral surfaces. There is a limit to the effect, however, determined by the maximum possible rate of extraction of the ions from the mineral lattice. In some cases, flow rate actually modifies chemical process. Felspar hydrolysis, for example, proceeds beyond the clay mineral stage to an insoluble accumulation of aluminium hydroxide (bauxite) in rapid-flow (intense leaching) environments with the loss of all silica in solution.

Chemical environment is clearly very complex and depends on the reagents present. In practice, though there are two chemical parameters which adequately describe chemical environment, pH and Eh. pH or *acidity* is a measure of hydrogen ion concentration in a solution and is expressed as the negative logarthim of the hydrogen ion concentration. In pure water at 25 °C, $[H^+]$ is 10^{-7} moles/litre, and the pH is 7. Very strong acids approach a pH of 0, and strong bases a pH of 14, but natural weathering environments rarely exceed the range of 4 to 9. Eh or *oxidizing potential* is a measure of the ability of a solution to transfer electrons to and from ions. (It should be remembered that oxidation is a process of removing electrons from ions and that reduction is the reverse.) Eh is measured by comparing the electrical potential necessary for a red-ox reaction with the reduction of hydrogen ions under standard conditions, which is allocated an arbitrary Eh value of 0 volts. Positive Eh indicates oxidation while negative values indicate the reverse, reduction, process. These parameters are very important since almost all weathering reactions involve the loss or gain of H^+ or OH^- ions, the loss or gain of electrons from ions, or both. Thus, changes in either or both of these factors influence almost all reactions. In fact, most weathering systems can be characterized by graphs of pH against Eh alone. As

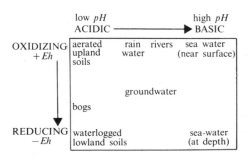

FIG. 1.11. Some weathering environments defined chemically.

examples we have already seen that hydrolysis causes OH^- ions to be expelled and is therefore favoured in acid environments, while oxidation and reduction reactions are, by definition, controlled by Eh. Relationships between pH and Eh and some common weathering environments are shown in Fig. 1.11.

Further Reading

GARRELS, R. M. and CHRIST, C. L. (1965). *Solutions, Minerals and Equilibria*, Harper and Row, New York.

HARVEY, D. (1969). *Explanation in Geography*, McGraw-Hill, New York.

KITTEL, C., KNIGHT, W. D. and RUDERMAN, M. A. (1965). *Mechanics*, Berkeley Physics Course: vol. 1, McGraw-Hill, New York.

KRAUSKOPF, K. (1967). *Introduction to Geochemistry*, McGraw-Hill, New York.

PAULING, L. and PAULING, P. (1975). *Chemistry*, Freeman, New York.

2 Sediment in Transfer Systems—Its Nature and Properties

2.1. The properties of sediments and sedimentary particles

In chapter 1, concerned with the mechanics and chemistry of transport, sediment was implicitly treated as an inert, amorphous substance passively transported from place to place by the processes acting upon it. Clearly, this is a simplified view of sediment and of the role it plays in determining what processes are possible in any sediment transport system. The term 'sediment' may apply equally to massive boulders on a scree slope as to the finest clay particles sedimenting from the still waters of a lake, and yet these two materials have little in common. It is important to be able to define more closely the materials in which we are interested, not only for reliable description but also because sediment character frequently determines (or at least modifies) the type of transport process operating. For example, certain characteristics determine how a sediment will behave under a stress (section 2.3) such as that due to the downslope component of its own weight. In this sense sediment is able, by feedback mechanisms, to influence process and so cannot be considered entirely passive.

It follows that, if a sediment can modify the process acting on itself, its character to some extent indicates the process environment in which it was deposited. For these relationships between sediment character and process of transport to be of value in interpreting deposits, it is necessary to refer to a number of reliable sediment properties which may be used in description and comparison. It is also important to bear in mind the degree of alteration experienced by sediment in transport. During their progress through the system, sediment particles interact with the transporting media, the environment and other sedimentary particles to become sorted, reduced in size, altered in composition and so on, until they bear little resemblance to sedimentary particles derived by weathering processes. The way in which their character changes is an indication of how the system modifies with distance, particularly with respect to transporting power and erosive capacity.

The most useful characteristics for sediment description and relating sediment type to process environment are shown in Table 2.1. This is not an exhaustive list of sediment properties, but it includes those which have important relationships with transport-process, whether causally or whether as indicators of certain general environments.

Sediment properties can be classified into those applicable to single grains and to aggregates of grains alike, and those which are relevant only to a collection of grains (Table 2.1). Similarly, some characteristics are valid whether the sediment is in transport (dynamic case) or deposited (static case), and it is largely those properties which are possessed by single grains which remain relevant in both conditions. Collective properties are generally only relevant to deposited sediment. Out of the list of properties in Table 2.1, strength, itself a function of other properties, warrants special attention. The strength of a sediment is its resistance to movement and is therefore the deciding factor in determining the response to an applied force, and to some extent in controlling the possible range of processes which can operate. In

Table 2.1

Sediment characteristics

	Population and/or individual grain characteristic	*Dynamic and/or static characteristic*
Grainsize	P + I	D + S
Grain shape	P + I	D + S
Grain surface character	P + I	D + S
Composition	P + I	D + S
Porosity	P	S
Permeability	P	S
Sediment structure and fabric	P	S
Strength	P + I	D + S

the following section a brief examination of sedimentary properties is made, and in section 2.2 strength and its interrelationship with other properties are considered in more detail.

Grainsize

The size of a sedimentary particle is one of its most fundamental properties since it is in some way related to the force necessary to move it. Superficially grainsize is a simple concept but, as one examines the assumptions made in its measurement more carefully, it becomes apparent that the high variability in natural particle shape makes the meaning of size difficult to define. The most meaningful measure of size is particle volume, since it expresses the amount of space occupied by the particle and is directly relatable to weight, which we have already seen is important in analysing forces acting on a particle. Measuring grainsize by volume suffers from two major weaknesses; first, it is difficult to appreciate the meaning of a three-dimensional measure and secondly, volume is not practicably measurable for particles smaller than about a centimetre in diameter. The next best approach is to assume that sedimentary particles approximate in shape to some more or less regular body whose volume can be defined by the measurement of one or more axes. If this can be achieved, the simple relationship between what one measures as grainsize and particle weight is not seriously impaired. Volume in regular bodies, such as a sphere or a cube, is easily related to only one axial measurement. The diameter of a sphere, for example, is sufficient to define its volume using the simple formula for spherical volume. As a body becomes less regular, more and more axes are required in order to express its volume adequately, though more than three are seldom considered for natural sediment particles.

Small particles, that is silt- and clay-sized particles, are usually measured by sedimentation techniques. The technique relies on a relationship between rate of settling and particle size called Stokes's Law. Stokes's Law strictly speaking holds only for spheres, and so is a questionable assumption when applied to natural particles. This is especially so for clay particles which are usually markedly platy. Intermediate grainsizes such as sand and gravel are generally measured by passing a sample of sediment through a stack of sieves whose mesh sizes become progressively smaller downwards. The distribution of sizes in the sample is given by the weight trapped on each sieve. It is possible for particles of very different volumes to be trapped on the same sieve since different shapes of particle may be restricted by the same mesh size. Large particles, of the coarse gravel to boulder size ranges, are measured individually using callipers or a tape, and up to three axes may be recorded. These axes are mutually orthogonal and are labelled the *a*- (long) *b*- (intermediate) and *c*- (short) axes. When three axes are measured it is possible to adopt a much less regular model for particle shape, which improves the relationship between measured axes and volume.

It is important that measurements of grainsize be relatable in a consistent and known way to particle volume for two reasons. First, if grainsize is to be considered a fundamental property, it should be (though in practice never is) unique, independent of any other parameter such as particle shape. Secondly, and this is more important, almost all theoretical derivations of grain movement by transporting media, such as the transport of particles in a fluid, assume spherical particles are being transported. If one of our objectives in studying transport processes is to attempt to relate theoretical deductions to observed transport process, we must be aware of and able to correct for the fact that natural particles are markedly and consistently aspherical. Although grainsize is a characteristic of each particle in a sediment, it is generally measured on a collective basis for all but the largest sizes of material. Thus, the sizes of a number of grains in a sample are measured, which allows a grainsize distribution curve to be constructed. The grainsize distribution curve gives a considerable amount of information about the sediment and the process which transported it. Mean grainsize for example, is important in deducing power/transport relationships for processes. The shape of the curve can also be described statistically and can give numerical values for the degree of sorting of the sediment, among other things. A well-sorted sediment has a narrow range of grainsize and vice versa, and degree of sorting gives an indication of process-type (chapter 7).

Grain shape

Grain shape and size together are the properties of a sedimentary particle which define its macroscopic appearance, and we have already discussed the close interrelationship between the two properties under the measurement of size. It is to be expected that shape and size determination suffer from similar conceptual difficulties, since they share the necessity for comparison of particles with some standard reference shape. A distinction is made in the concept of shape between morphology and surface roughness of grains, which are logically independent properties. The most commonly used measure of morphology is grain sphericity which is, as the term suggests, the closeness to which shape approximates to a sphere. Sphericity is a function of the original grain shape and can be related to a number of factors; crystal form, perhaps, in the case of individual mineral fragments derived from an igneous rock. Subsequent wear due to attrition in transport tends to cause only slight modification of overall morphology. Surface roughness of a sedimentary particle is often referred to as its roundness and is an indication of the sharpness of its edges and corners. Roundness, unlike sphericity, is strongly related to the amount of wear suffered during processes of transport though, naturally, the initial particle roundness must be taken into consideration. On the small scale, roughness elements become difficult to distinguish from surface textural features and indeed the division between the two is essentially arbitrary.

Shape is not only related to wear during transport; it also influences the susceptibility of particles to transport. Generally speaking, the more spherical a particle is the easier it is to transport because only rolling friction need be overcome. Also, particles with lower surface roughness experience reduced sliding friction in motion.

As we have already seen, an exact description of particle shape is very difficult to achieve because of the extreme irregularity of natural grains. More detailed discussion of the conceptual problems involved may be found in the references in 'Further Reading' and will not be elaborated on here. Suffice it to say that most measures of shape currently employed are practical compromises which, while not adequately describing shape, do provide some basis for comparison between sedimentary materials and do show at least some relationship to the way materials behave.

Grain surface features

The surface textural characteristics of a particle are only arbitrarily separated from roughness elements at one level, though on a microscopic scale differentiation between surface features has become extremely sophisticated. It is only recently that techniques have been developed to examine grain surfaces adequately, since the resolving power of conventional optical microscopes is rather coarse. In recent years, however, scanning electron microscopes have been adopted as the standard instrument in examining grain surfaces. Grain surface features have been used with qualified success to differentiate between sedimentary environments. From the point of view of process, grain surface characteristics are an extension of grain roughness and probably play a part in determining grain to grain friction.

Composition

No description of a sediment is complete without a compositional analysis. Small grains tend to be monomineralic, and identification is then a mineralogic exercise; larger grains tend to be rock fragments requiring petrologic description. Composition is the clue to provenance, that is the source area for the clasts contained in the sediment. In the geologic column this is often exceedingly difficult to do because frequently the source rocks no longer exist or outcrop, but in studying present-day sediments this problem is much less important. Consequently, compositional analysis of a sediment can tell us the derivation of the material and also the relative persistence of different rocktypes in transport processes. What part does composition have to play in the way processes operate and does it exert any control over sediment strength? Inasmuch as particle density, and hence weight, is controlled by mineralogy, composition undoubtedly influences the relative ease with which material is entrained in fluid flow and maintained in suspension. But since the range of density in commonly occurring natural particles is rather small, this is a minor effect. Perhaps the most important contribution composition has to make in the control of transport processes is through its influence on sediment strength. There is a broad distinction between clay and non-clay minerals, the former possessing significant interparticle cohesion and lower grain to grain friction. There are, however, no significant variations in strength parameters between the non-clay minerals, except at the level of intact rock.

Porosity

Porosity, a property of a static assemblage of particles, is not a constant for a given material but depends on the way in which it has been deposited, the amount of disturbance suffered and so on. It is defined as the ratio of pore-space volume to total volume of the sediment (V_p/V), though another measure of fractional pore-space, called the void ratio and defined as the ratio of pore-space volume to solid volume (V_p/V_s), is used by engineers. A simple equation relates porosity and void ratio:

$$\text{porosity} = p = \frac{(\text{void ratio})}{(1 + \text{void ratio})} = \frac{e}{1 + e}$$

It is often necessary to distinguish between primary porosity, or the porosity due to intergranular pore-space, and secondary porosity which is due to cracks, fissures, animal burrows and other large secondary voids within the material. In many instances the secondary porosity is far greater than the primary which, in the case of unweathered intrusive igneous rocks, may be very low indeed.

Since porosity is a property of a grain aggregate its value is strongly related to the nature of the individual grains and to the way they are arranged with respect to one another. The porosity of an assemblage of equally sized spheres, for example, may vary from a minimum of 0·26 in the closest packing array to a maximum of 0·48 for the loosest possible regular packing arrangement. Grainsize distribution also has an influence on porosity, especially if small grainsizes are able to fill the pore-spaces between the large particles. Finally, particle shape has an important part to play in the way grains can be packed together and therefore is a significant factor in the porosity of a sediment.

Porosity is a measure of the closeness of contact between particles and may therefore be expected to have some influence on how well they interlock with one another. Interlocking is in turn an important factor in the strength of grain to grain contact and, therefore, in the ease with which particles may be moved. Extremely closely packed particles are more difficult to pick up in a fluid flow for example and a considerable amount of work has shown that the internal strength of a sediment (ability to resist internal shear or sliding under stress) is strongly related to porosity (section 2.3).

Permeability

Permeability, like porosity, is an aggregate property concerned with the pore-space between the grains. It is a measure of the ability of the pore-network to transmit water and once again the distinction between primary and secondary permeability is valid. Many indices of permeability have been proposed of which we shall discuss only the saturated hydraulic conductivity, which applies to saturated flow conditions through porous media. We may define saturated hydraulic conductivity from Darcy's Law, in which:

$$\text{coefficient of hydraulic conductivity, } K = Q/IA$$

where Q is the discharge of water through a sample of cross sectional area A under a hydraulic gradient of I. A simple illustration of hydraulic gradient is shown in Fig. 2.1, for the case of a water table or phreatic surface beneath a sloping surface. The total head-difference between any two points is shown by the rest level in piezometers or standpipes inserted at those points. Darcy's Law is an empirical relationship which was derived from experimental flow data through a column of soil. The law applies only to laminar flow conditions, that is flow in which the stream lines are parallel and show no tendency to turbulent eddying (section 6.1) and consequently it does not explain flow through very large pore-spaces.

Hydraulic conductivity is closely related to grainsize. Fig. 2.2 shows a good logarithmic relationship which only begins to break down for grainsizes larger than about 3 mm, when turbulent flow starts to occur. The graph shows that a 10 fold increase in grainsize causes 100 fold increase in hydraulic conductivity. The results are for uniform sized spheres, but in

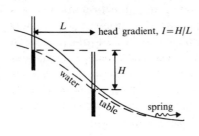

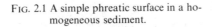

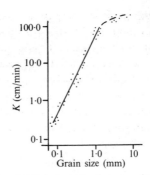

FIG. 2.2 Relationship between hydraulic
conductivity (K) and grainsize for
spherical grains.

FIG. 2.1 A simple phreatic surface in a homogeneous sediment.

materials with a distribution of grainsizes it is probably the smaller particles which have most influence over permeability. In fact, this is well shown by an established empirical relationship between hydraulic conductivity (K) of sands and effective grainsize called the Hazen formula:

$$K = 100 \, D_{10}{}^2$$

where D_{10} is the grainsize of which 10 per cent is smaller (10th percentile). This demonstrates that smaller grainsizes have a disproportionately large effect on hydraulic conductivity. Many laboratory experiments have shown a similar effect. Porosity and pore-space geometry, as well as other para- meters which influence them (fabric, packing, particle shape) have a signifi- cant influence on permeability, though the effect is generally not so marked as for grainsize. It has been suggested that an increase in porosity from 30 to 40 per cent causes a concomitant increase in hydraulic conductivity of about $2\frac{1}{2}$ times. Clay soils, which can sometimes show a very wide range of void ratio, may demonstrate somewhat greater variations in hydraulic con- ductivity.

The ability of a sediment to transmit water affects its strength through its influence on pore-water pressures. Obviously, the slower that water is able to pass through a material, the higher is the probability of pore-water pres- sure developing within it. We have already mentioned the effectiveness of pore-water pressures in relieving normal stress between particles in section 1.5. A reduction in normal stress, caused by buoyant upthrust, diminishes interparticle frictional resistance within the sediment, and hence total strength decreases. This is an extremely important aspect of the strength of aggregate of particles (chapter 4).

Sediment structure and fabric
Sediment particles at rest and in motion tend to organize themselves into forms which reflect the way they were deposited or the way in which they are being moved. Thus the sediment bed of a river is frequently convoluted into bedforms which characterize the water flow forces exerted at the bed.

Bedforms are frequently preserved in geologic formations where they can be used as valuable indicators of flow regime and current direction in interpreting past environments. From the sediment transport point of view, their interest lies in their relationship with the mechanics of fluid movement. In many environments sediment particles tend to orientate themselves so that they present least resistance to motion. The collective orientation of particles in a sediment is called its *fabric* and is an indication of the way particle movement took place. For example, sliding particles tend to orientate themselves with their long-axes parallel to the transport direction, while rolling particles are usually transverse.

2.2. Strength of sediments—their ability to resist transport

The strength of a sediment can be defined loosely as its maximum resistance to applied stress. Stress may, for the time being, be regarded as force per unit area (for example kg/m^2) though its meaning is expanded in section 2.3. The concept of strength is equally applicable to the movement of individual particles as to aggregates of grains in contact with each other, and it is influenced to a greater or lesser extent by many of the properties discussed in section 2.1. It is worth mentioning at this point that, since many sediment properties are not constant for a given material, strength itself is not a material constant. Sediment properties influence maximum strength and also behaviour under stress (section 2.3) but it should not be assumed that they are the only important factors in strength. The direction and rate of stress application, length of time over which it is applied, and the previous history of movement in the sediment may all have a part to play in determining strength at any given time. To summarize: sediment strength is often as much related to stress factors as to the properties of the material itself.

The components of strength: friction and cohesion

Only two sources of resistance to motion need to be considered, namely friction and cohesion, and these may be defined by their relationship with normal stress. Friction is an inter-surface resistance to shear (displacement along the surface) which is proportional to the normal stress pushing the surfaces together. Frictional resistance has already been mentioned in 1.4 and you may recall that it exists between sliding or rolling bodies on a surface. We shall not consider the mechanisms responsible for friction here but a brief and simple account may be found in Lambe and Whitman (1969, chapter 6). Recall also that frictional resistance is proportional to the applied shear stress (stress tending to cause shear), so that if no shear stress is applied to a body then no frictional resistance to motion is mobilized. Hence a stationary particle on a horizontal surface, as in Fig. 1.6, has no stress applied to it tending to cause lateral motion and no frictional resistance to motion. As increasing lateral stress is applied to the body, frictional resistance builds up in the contact area to resist it until a maximum value is reached. When the maximum possible frictional resistance is exceeded by a shear stress, the body will begin to accelerate along the surface. The maximum amount of frictional resistance which can be mobilized depends on normal stress and on a constant of the contact surface called the coefficient of plane sliding friction. Figs. 1.6 and 1.7 illustrate two very frequently encountered force situations in loose particle transport; Fig. 1.6 where shear stress is due to a transport agent such as fluid flow and 1.7 where the stress is derived from the particle's own downslope weight component.

Cohesion can be defined as resistance to shear which is independent of normal stress across the surfaces. A measurable resistance to motion exists even when there is no normal force pushing the surfaces together. Cohesion is, then, analogous to a glue which has a certain strength no matter what normal stresses are applied across the surfaces. It can develop in a number of ways in quite different materials. The weakest of all cohesional bonds occur between very small (silt- and clay-sized) sedimentary particles composed of *non-clay* minerals, for example, quartz and felspar. These particles are uncharged but a certain amount of cohesion exists between them due to van der Waals' forces, though since the bonds are quite weak they have a negligible influence over larger and therefore heavier particles. Smalley (1972) calls these bonds *inactive* because for all practical purposes they do not operate once interparticle contact has been broken. Clay minerals on the other hand, exhibit repulsion and attraction forces between particles due to their negative charge, polarized pore water, and the presence of solute ions in the pore water. The nature of the forces is extremely complex and whether there is a net tendency towards attraction or repulsion between particles depends largely on the quality of the pore-water. Basically, attraction between particles is due to negatively charged clay particles attracting positive solute ions and water on to their surfaces to form a strongly bonded solid-fluid system. Smalley (op. cit.) calls the cohesive bonds between clay minerals *active*, because they are able to operate over fairly large particle separations. Consequently, clay cohesion is not entirely lost when the particles are disturbed, and thus it is frequently a very important part of total clay strength. Some cohesion may result from negative pore-water pressures which tend to draw grains together by a suction effect. *Capillary cohesion* (section 1.5) seldom becomes an important factor in strength because negative pore pressures are usually short-lived on geomorphic time-scales. By far the highest cohesive forces occur between the constituent grains and minerals in consolidated sedimentary and igneous rocks. During lithification, sedimentary rocks become cemented by mineral cements, derived internally by pressure solution at grain contacts, or transferred into the sediment by groundwater circulation. The most frequently encountered cements in sandstones are quartz and calcite, and they may be as strong as or stronger than the grains themselves. The constituent mineral grains of igneous rocks are, of course, fused together on cooling, and consequently the cohesive strength of the intact rock is so great that frictional resistance to shear is of minor importance.

We shall now consider specific examples of sediment strength, dealing first with the simple situation of loose particles on a surface and leading on to aspects of aggregate strength. This distinction between individual grains and aggregates is a convenient one because transfer processes can be broadly distinguished according to whether they shift sediment as discrete particles or in bulk.

Resistance of single grains to motion

The frictional resistance of individual grains to motion has been considered in section 1.4, but it will be useful to re-examine it here. Let us then, consider a single particle resting on a surface of angle α and subjected to a force F. The surface could be a river bed or a hillslope and the force might be derived from any external influence such as fluid flow (Fig. 2.3a). This may be considered as a general case for the initiation of individual particle transport. The *stresses* tending to cause movement are F/A and $W \sin \alpha/A$; while those resisting motion are $(W \cos \alpha/A) . \tan \varphi_{\mu s}$ and c/A, where c is the cohesive force between the surface and particle, $\varphi_{\mu s}$ is the static angle of plane sliding friction (section 1.4) and A is the area of contact between surface and particle. At *failure* (the point when movement begins) the stresses causing motion are just balanced by those resisting it and

$$F/A + W \sin \alpha/A = W \cos \alpha . \tan \varphi_{\mu s}/A + c/A.$$

Cancelling out A and simplifying we have

$$F = W \cos \alpha . \tan \varphi_{\mu s} - W \sin \alpha + c.$$

For a non-cohesive surface, this simplifies further to

$$F = W (\tan \varphi_{\mu s} - \tan \alpha).$$

This formula shows that, for a non-cohesive surface, the external force required to cause particle sliding becomes zero when α is equal to $\varphi_{\mu s}$ (section 1.4 and Figs. 1.6 and 1.7). It should be noted that these formulae apply only to situations where force F is applied parallel to the slope. If a force (L) causes an uplift on a particle, and this is possible in fluids due to aerodynamic lift and upward turbulent eddying (E), then grains may be picked from the bed. When this occurs tension is set up between grain and surface, therefore reducing frictional resistance to zero, and the only forces of re-

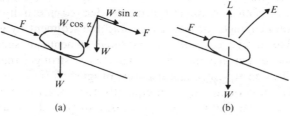

(a)   (b)

FIG. 2.3. Resistance of single grains to transport.

sistance to be overcome are cohesion and the submerged grain weight (Fig. 2.3b). This example illustrates well how strength is related to the direction in which force is applied, and that frictional resistance does not operate across surfaces in tension.

Referring again to Fig. 1.6, recall that a body on a frictional surface will begin to slide when

$$F = R \tan \varphi_{\mu s}$$

where R is the normal reaction (weight of body) and F is the applied force. If F is plotted against R for a number of bodies of different weight but the same $\varphi_{\mu s}$, we obtain a straight-line relationship through the origin, and at an angle $\varphi_{\mu s}$ to the horizontal axis. Fig. 2.4 shows the relationship; since maximum resistance to shear (shear strength) is equal to applied shear stress at failure, the diagram can be taken as a strength model. Thus, for a cohesionless contact, in stress terms rather than total forces:

$$\text{shear strength} = s = \tau_f = \sigma \, . \tan \varphi_{\mu s}$$

where τ_f is the shear stress at failure and σ is the normal stress (Fig. 2.4). If cohesion exists between grain and surface, then the graph no longer passes

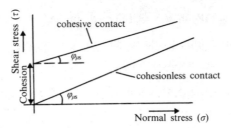

FIG. 2.4. Relationship between applied force and normal stress for a sliding contact.

through the origin but makes an intercept equal to the cohesion with the vertical axis. The relationship for strength becomes:

$$s = \tau_f = c + \sigma \tan \varphi_{\mu s}$$

which is well known in soil mechanics as *Coulomb's failure law*, and is a general model of material strength. Note that it assumes cohesion to be additive and friction to be linearly related to normal stress.

So far we have considered frictional resistance only for static contacts. Resistance still exists when motion is taking place though often it tends to be less, which explains why it is more difficult to get an object moving than to maintain its motion. Frictional strength then, reaches a peak value immediately prior to failure and declines to a lower level once motion begins. We can, therefore, think of any contact as possessing two angles of sliding friction; a static angle ($\varphi_{\mu s}$) and a dynamic angle ($\varphi_{\mu d}$). This loss of strength is reversible; that is to say, if the body stops it will recover its static frictional strength. We shall discover below that strength loss due to shearing is not always fully recovered in the case of aggregate strength.

Resistance of grain aggregates to shear

One of the ways in which the strength of an aggregate may be determined is by the direct shearbox test and it will be helpful to the understanding of inter-particle resistance to look at this test. The shearbox apparatus (Fig. 2.5) consists of a split metal box into which a sample is placed, with a mechanism for pushing the lower half of the box and a load-measuring device called a proving ring on the upper part. A normal load is applied to the sample through a weighting system, and the lower half of the box is driven forward

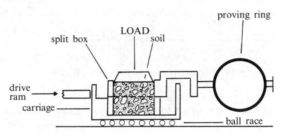

FIG. 2.5. The direct shearbox.

by a motorized ram, while movement in the top half of the box is resisted by the proving ring. The specimen is then driven forward until it shears along the two box halves. The maximum proving ring load measures the shear stress at failure and if many tests are carried out at different normal loads, a graph of normal load against shear load can be plotted. The resulting graph is exactly similar in principle to that for plane sliding friction (Fig. 2.4) except that the angle made with the horizontal is the *angle of internal shearing resistance* (φ). In the case of an aggregate, shear takes place on a line between grains rather than between a particle and a surface. The general strength model is, therefore, very similar to that for a single sliding grain:

$$s = \tau_f = c + \sigma \tan \varphi.$$

Two points must be remembered when considering the shear strength of aggregates. First, in addition to plane friction and cohesion, a source of resistance is derived from particle-interlocking, which always gives rise to a non-planar surface upon which shear takes place. Secondly, the shear surface is not well defined, as is the case for a single grain on a solid plane, but can shift its position in the material as movement progresses. Coulomb's failure law above is an empirical fit to observations and suggests that granular materials effectively behave as if shear were taking place on one simple plane through the medium. It should be remembered that this is only an empirical fit to a complex process.

It is impossible for a perfectly planar shear plane to pass through a granular medium unless grain splitting occurs, due to the interlocking of particles. As an example in the regular packing of cohesionless spheres shown in Fig. 2.6, the top layer of particles will not begin to slide until the tangents of the grain contacts $(A-A')$ are inclined at $\varphi_{\mu s}$, requiring an over-all slope angle of $(30 + \varphi_{\mu s})°$. Naturally, this example is somewhat artificial because packing and grain shape as regular as this would never occur naturally, but it does serve to demonstrate the point. Effectively, those grains above the shear plane must be lifted bodily over those below in order to move. The average

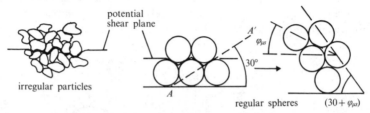

FIG. 2.6. Shearplanes in granular materials.

angle through which grains are lifted on the shear plane is called the *dilatation angle* (θ), which is a part of the angle of internal shearing resistance:

$$\varphi = (\varphi_{\mu s} + \theta).$$

Dilatation angle, and therefore the angle of internal shearing resistance (φ), is related to the degree of grain interlocking, a function of packing and porosity of a sediment. One might expect a relationship between peak strength and porosity, and Fig. 2.7 demonstrates this for a number of materials. The effect is quite significant; for example, a drop of 8–10° in angle of internal shearing resistance results from an increase in porosity of about 10 per cent for the materials shown in Fig. 2.7.

One of the objections to measuring strength in a shearbox is that the sample is forced to shear at a fixed rate, rather than to move freely once failure has occurred as it would do in a natural landslide. This objection can be overcome for cohesionless dry aggregates by measuring the angle at which a pile of the material will begin to avalanche. There are many methods of achieving this; all involve tipping a pile of material until sliding occurs. Two critical angles are observable, an upper angle when avalanching begins and a lower one at which slipped material comes to rest. Both of these angles have, at some time, been called *the angle of repose* and so it seems safer to adopt the terminology used by Allen (1969), who called the upper angle the *angle of initial yield* (φ_i) and the lower angle the *angle of residual shear* (φ_r). The angle of initial yield tends to be rather variable and depends, like the angle of internal shearing resistance, on the degree of particle interlocking before shearing takes place. It seems reasonable to assume that φ_i is roughly equivalent to φ, measured in a shearbox test, since both are measures of frictional strength before failure and are influenced by in-

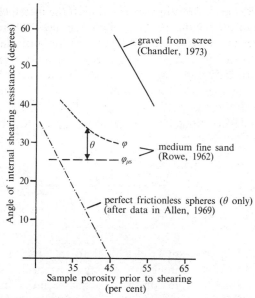

FIG. 2.7. The effect of porosity on angle of internal shearing resistance.

terlocking. The angle of residual shear (φ_r) is less easily related to shearbox test measurements, because the sample behaves very differently after failure. The slipped material is free to avalanche in the angle of repose test, but is constrained to move at a slow rate in a shearbox. Nevertheless, results seem to suggest that φ_r is at least roughly comparable with the angle of internal shearing resistance of the loosest packed state. The difference between φ_i and φ_r represents a loss of strength due to failure. When the material being tested is in its loosest packed condition, the strength loss is recoverable; that is to say, if repeated testing is carried out there is no tendency for φ_i to decline. For a loose material then, the difference between the two angles is probably analogous to the difference between static and dynamic plane-sliding friction angles ($\varphi_{\mu s}$ and $\varphi_{\mu d}$). If the sample initially has a low porosity, some part of its strength is not recoverable after failure, because it undergoes a non-reversible increase in volume and porosity as shear takes place.

The dependency of φ_i on porosity introduces the possibility that some aspects of strength can be lost through failure and not recovered when shearing ceases. This lower value of strength in disturbed materials is referred to as *residual strength* and it can be understood more fully by returning to the shearbox text. A densely packed sand, for instance, dilates during the shear test. If shearing load is plotted against displacement in the specimen (Fig. 2.8) the sand reaches a peak strength immediately prior to failure, which falls to a constant lower level as the sample ceases to dilate. The loss in strength is not recoverable since it is caused by an increase in porosity. Loose-packed sand, on the other hand, tends to decrease slightly in volume due to shear, and instead of reaching a peak value its strength rises more

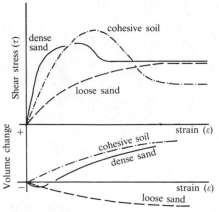

FIG. 2.8. The effect of shearing on sediment strength.

gradually to reach a constant level similar to the residual strength of the dense material. The angle of internal shearing resistance of a sediment in the state of no net volume change during shear has been denoted φ_{cv} (the angle of internal shearing resistance at constant volume; Bishop, 1966). It is roughly a constant for any given material and is related to a standard loose-packed condition. It should not, however, be assumed that no volume changes are taking place when a loose-packed material is sheared, since particles above the shear plane must still ride up and down over those below. Upward and downward movements balance each other, so that there is no *net* change in sample volume as shear continues.

A further loss of strength tends to occur in cohesive media, due to the tendency for cohesion to be lost after failure has taken place. In some cases, where cohesion was a major proportion of the intact strength, residual strength may be only a small fraction of the undisturbed strength (Fig. 2.8). Residual strength is a very important factor in landslide processes and is reconsidered in chapter 4. Another important aspect of sediment strength which has been postponed to chapter 4 is the part to be played by pore-water pressures. So far we have only been concerned with dry strength, but we shall see later that the presence of pore–water can have a dramatic effect on the effective strength of a granular material.

Some confusion may result from the numerous different angles of sliding friction which have been introduced in this section. They are each related to a different condition of the sediment or a different type of test procedure and they are summarized in Table 2.2. Reference to this table may prove useful in this and other chapters.

Some relationships between sediment properties and strength

We have already mentioned that our main interest in sediment properties lies in the way they influence transfer processes through their control on strength. Some of the more important linkages between properties and

Table 2.2

Angles of shearing resistance

Symbol	Definition	Comments
$\varphi_{\mu s}$	Static angle of plane sliding friction	Angle at which a loose particle begins to slide down surface. Approx. constant for a given particle: surface contact.
$\varphi_{\mu d}$	Dynamic angle of plane sliding friction	Angle at which loose particle *just* comes to rest. Approx. constant.
φ, φ'	Angle of internal shearing resistance	Angle at which shearing begins in loose aggregate. Measured in triaxial or shearbox test (section 2.3). Not a constant; depends on initial porosity. ' Denotes pore-water pressures considered (See chapter 4).
$\varphi_{cv}, \varphi'_{cv}$	Angle of internal shearing resistance (constant volume)	Angle at which shearing continues to occur after peak strength is passed and no further volume change takes place in material. Approx. constant for a given material. Refers to a test in triaxial or shearbox.
φ_i	Angle of initial yield	Angle at which cohesionless aggregate begins to avalanche. Not a constant, depends on porosity.
φ_r	Angle of residual shear	Angle at which cohesionless aggregate comes to rest after shearing. Approx. constant for a given material.
$\varphi_{res}, \varphi'_{res}$	Residual angle of internal shearing resistance	Angle of internal shearing resistance for a material which has undergone considerable shear. Approx. constant for a given material.

strength parameters are summarized in Fig. 2.9, and it is apparent that the influence may be direct, or may operate through control of another sediment property. In particular, aggregate properties are largely responsible for the difference between peak and residual frictional strength, though single grain properties do have some role to play in this respect through their influence upon aggregate properties. In addition, one may also generalize that grain-

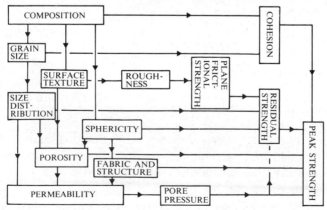

FIG. 2.9. Inter-relationships between sediment properties and strength parameters.

size and composition exert the most powerful influence on cohesion, and that shape factors exercise the greatest control upon residual strength. Attention is also drawn to the links between permeability, pore–pressures and strength parameters which are dealt with in chapter 4. The relationship between composition, grainsize and cohesive strength has already been alluded to, in that active cohesive bonds are peculiar to clay minerals while shorter range inactive bonds are found in fine-grained non-clay materials. With regard to frictional strength, reference has been made to the relationship between porosity and angle of internal shearing resistance (Fig. 2.7).

Turning now to some unconsidered relationships between strength and properties, the effect of grainsize is equivocal. Many workers have reported conflicting relationships between grainsize and shear strength but in view of the conflict, and the fact that there is no geometric reason for grainsize to have any influence on strength, it is probable that variations in other parameters have been responsible for the confusing relationships in the literature. *Grainsize distribution*, in contrast, can have a marked effect on strength. It is difficult to study its influence rigorously but we can take as an example the variation of coarse- to fine-grained material in a simple two-sized mixture. Results of experiments on mixtures of grainsizes (Fig. 2.10) show that fric-

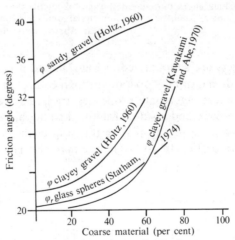

FIG. 2.10. The effect of grainsize distribution on strength.

tion angle (φ or φ_r according to the experiment) increases markedly with increasing coarse-content.

The effect of *grain surface roughness* on frictional strength is best summarized by Table 2.3. The results of many studies have demonstrated that greater angularity is accompanied by an increase in friction angle. Note also the effect of metal tarnish on the friction angle of lead and steel spheres. Clearly, greater surface roughness is responsible for increasing the amount of surface-to-surface interlocking and therefore the frictional strength.

Table 2.3

Angles of repose for natural and artificial materials

	φ_i	φ_r
Steel balls { rusty	25·3°	21°
untarnished	33·8°	28·1°
Lead shot { graphite coated	22·5°	21·2°
oxide coated	28·0°	24·6°
Smooth glass spheres		20–25°
Natural dune sands		30–33°
Subangular sands		31–35°
Angular gravel and talus		38–42°

Source: Based on data from Allen (1970), Simons (1956), Carrigy (1970), Chandler (1973), Blight (1969), Statham (1974).

Overall grain shape also undoubtedly influences the friction angle, but there have been very few studies aimed directly at this problem.

Only a few relationships between sediment strength and properties have been introduced here. How sediment character affects the operation of specific processes is postponed to the relevant chapters.

2.3. Models of material behaviour under stress

The meaning of stress

Up to this point a working definition of stress as force/unit area has been sufficient for our purposes, though it will now be useful to extend the concept a little further. First, our simple definition assumes that stress and pressure are analogous, which is not precisely the case. For example, the atmosphere exerts a pressure of about 1 kg/cm² on the earth's surface and in turn *stresses* are set up within the earth's surface to resist that pressure. It is the internal force of resistance per unit area which should be regarded as a stress, not the externally applied *pressure*. In practice, however, they are often numerically equal.

It will be useful in the understanding of stresses in sediments to consider another test of sediment strength, *the triaxial test*. The apparatus is illustrated in Fig. 2.11. A cylindrical specimen, enclosed in a watertight rubber membrane within the cell, is subjected to a confining pressure through the cell fluid and to a vertical load through a drive system and proving ring. Vertical load is increased by driving the specimen upwards on to the proving ring, at a fixed rate, while cell pressure is held constant. The stresses are applied so that no shear stresses exist on the sample surface, only normal stresses. Consequently, the stresses on the sides and ends of the sample are called *principal stresses*, because they are stress directions which have no component of shear. At the start of the test, cell pressure acts over the whole sample and so vertical and horizontal principal stresses are equal.

Throughout the test, load is applied to the specimen ends so that the

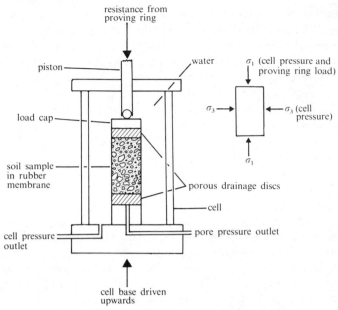

FIG. 2.11. The triaxial apparatus.

major principal stress (σ_1) becomes steadily greater than the minor principal stress (σ_3), which is equal to the cell pressure. As σ_1 increases the sample deforms and shortens in length, or alternatively it becomes *strained*. Strain (ε) is defined as:

$$\varepsilon = (d_l/l_o) \times 100\%$$

where d_l is the change in the length of a sample of original length l_o. Eventually the resistance of the specimen to applied stress is exceeded and it shears along a plane, inclined to the principal stress directions, upon which shear stress exceeds shear strength. Obviously this plane must be inclined to the principal stress directions for shear stress to exist upon it. Values for cohesion and internal shearing resistance can be obtained from a number of tests on a material at different confining pressures. The results are plotted on a graph of normal stress against shear stress (Fig. 2.12a). The stresses at failure on the planes normal to σ_1 and σ_3 plot on the horizontal axis, since no shear stress exists on these planes (*principal planes*). A circle drawn through these points, whose diameter equals ($\sigma_1 - \sigma_3$), defines the shear and normal stresses existing on any inclined plane passing through the sample. This is the *Mohr Stress Circle* and those who wish to understand its usage in anything other than a qualitative sense are referred to the excellent account in Carson (1971a, chapter 1). We can appreciate the operation of Mohr Circles from Fig. 2.12a. Points X and Y represent respectively the stress conditions on planes parallel and normal to the major principal stress direction and, as expected, only normal stresses exist on these two planes. The Mohr circle describes the state of stress on any plane, such that point A gives the normal and shear stresses existing on a plane making an angle of α with the major principal plane (normal to σ_1) or $(90 - \alpha)°$ with the principal stress direction. From the diagram then, normal stress is equal to OB and shear stress to AB. Figure 2.12b shows Mohr circles at failure, plotted for a number of tests on the same material. Since these are all failure circles it follows that the tangent to them separates stress conditions into those not causing failure below the line and those which must cause failure above. The point at which the circle touches this *Failure Envelope* gives the state of stress and the inclination of the plane upon which failure first occurred in the test. Points A_1, A_2, etc. represent the ratio between normal and shear stress at failure and therefore, bearing in mind that shear strength at failure is equal to shear stress, angle of internal shearing resistance and cohesion may be interpeted from the failure envelope in the same way as for a direct shear test (Fig. 2.4). It can also be seen from the geometry of triangle A_2BC (Fig. 2.12b) that the failure plane must be inclined at an angle of $(45 + \varphi/2)$ to the major principal plane.

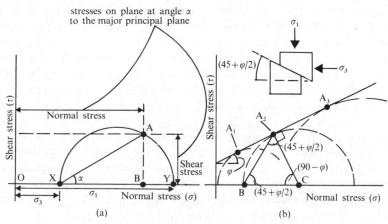

FIG. 2.12. Mohr stress circles and the failure envelope.

Idealized models of behaviour under stress

Behaviour under stress is defined by the relationships between applied stress, strain and time. There are three major models of behaviour which have been proposed: elastic, plastic and viscous deformation. They are entirely general models, referring to natural and artificial materials alike, and often explain deformations in artificial materials more successfully than in sediments.

Elastic deformation occurs when strain is proportional to applied stress. In the strictest sense, elasticity requires a linear relationship between stress and strain though some materials, for example rubber, demonstrate a non-linear relationship. The important aspect of elastic behaviour is that all strain disappears when applied stress is removed. A simple model of an elastic material is provided by an ordinary spring, in which the degree of extension or compression is proportional to applied load. Its original length is regained when the load is removed except that beyond a critical level of stress, a yield point called the *elastic limit*, elastic behaviour gives way to failure. Inter-molecular bonds in elastic materials must act as very small springs to explain this behaviour.

A material is said to be *rigid plastic* if it shows no tendency to deform below a critical level of stress or yield point. Consequently, no strain is experienced in the material as stress is applied until the yield point is reached and then strain occurs abruptly in the form of a failure. In contrast to elastic strain, plastic strain is irreversible on unloading. The movement of a rigid body on a frictional surface, to which we have had ample cause to refer, forms an ideal model of a plastic deformation. No movement occurs in the body as stress is applied until the yield stress is reached, at which point it begins to accelerate in the direction of applied stress. It is worth noting at this point, that failure along a well-defined plane is a characteristic of plastic materials since neither elastic nor viscous substances require a threshold level of stress for strain to begin. Consequently, mass movements such as those described in chapter 4 are indicative of plasticity in sediments, though this is not to say that sediments are in any way perfect plastic materials.

Viscosity is a property of fluids or fluid-like materials. In a viscous material the slightest amount of applied stress will cause strain to occur and to continue until stress declines to zero. In this sense then, strain is independent of stress though the strain *rate* is proportional to applied stress. True or Newtonian liquids show a linear relationship between stress and strain rate and the graph slope is the coefficient of viscosity. Non-Newtonian liquids may be pseudoplastic (shear thinning) if the rate of shear increases with applied stress, or dilatant (shear thickening) if the converse applies. Viscous behaviour is illustrated by the dashpot, a loose-fitting plunger in a fluid-filled cylinder. Any applied stress will push the plunger into the pot and force the fluid to escape round its sides, though greater stress will give a faster rate of movement.

The ideal models of deformation are illustrated in Fig. 2.13. Model stress-strain, strain-time and stress-strain rate curves are also shown. Combinations of the three types of behaviour are possible in the form of hybrid models. We shall now go on to look at the behaviour of natural sediments under stress and see how they compare with the ideal models described above.

Stress-strain behaviour of natural sediments

Let us begin by examining the behaviour of cohesionless materials, such as sand or gravel, when subjected to stress. Stress-strain curves from shearbox tests on loose-packed and densely-packed coarse sand are shown in Fig.

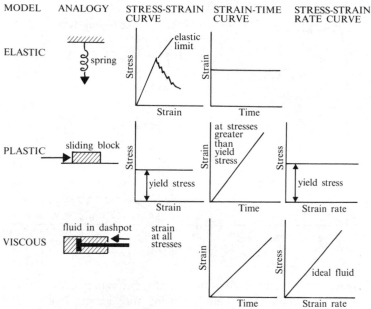

FIG. 2.13. Idealized models of material response to stress.

2.14. Notice first that both curves show a peak of stress before declining to a lower value, though this is only marked in the dense material. This is due to the difference between peak and residual strength, discussed earlier, and requires no further explanation here. Loose sand reaches its peak slightly later than coarse though both reach it in less than 3 per cent strain. Secondly, it is interesting to note that although strain is proportional to stress prior to failure, the curve is non-linear in this region in both curves. In fact, neither of the curves bears a particularly close resemblance to the model curves in Fig. 2.13. It is true that the sand demonstrates strain proportional to stress before failure, a characteristic of an elastic medium, but we would find that much of this pre-failure strain is irreversible on removal of the applied stress which is not characteristic of elastic strain. Furthermore, strain would cease to take place if applied stress were held constant below yield stress and so we cannot attribute the sand's behaviour to viscosity. The key to the stress-strain behaviour of a sand lies in its two-phase nature, that is to say in the fact that it is made up of solid grains and pore-spaces. We are only considering pore-spaces filled with air, but it must be remembered that water is often present, providing another complicating factor. As mentioned above, a small amount of strain is recovered on unloading and this may be attributed to elastic deformation of the grains themselves. We are still left with the problem of explaining the non-reversible but stress-dependent strain which occurs prior to failure. It seems likely that this strain is due to non-uniformity in the grain-packing, causing stress to be more concentrated at some grain contacts than at others. Certain contacts

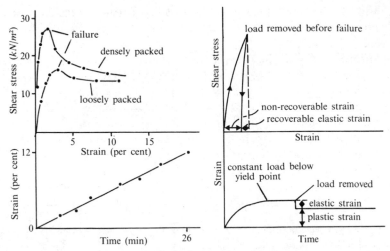

FIG. 2.14. Behaviour of a sand in a shearbox test.

are likely to fail before over-all failure occurs, and will generate grain sliding into a lower-stress position before ceasing movement. Thus, small local strains take place throughout the sample due to local grain sliding, and since these are plastic strains they are not regained on unloading. As applied stress increases, more and more contacts fail until general yielding occurs when peak strength is exceeded. Failure is a progressive process even in sand, to the extent that it does not behave as a perfect rigid plastic body before failure, though it should not be considered in any way analogous to progressive failure in clays (chapter 4). It must be remembered that all this takes place at very small strains. After failure, the strength of a sand falls rapidly to its residual level and subsequent deformation approximates to plastic sliding. Finally, since a sand exhibits no viscous tendencies, the rate at which strain occurs has very little effect on the form of the stress-strain curve itself.

Clay materials behave in a rather different way to sands, as we may deduce from the stress-strain curve in Fig. 2.15a. One major difference is that the clay only reaches peak strength after a considerable amount of strain, about 16 per cent in this case. A considerable amount of yielding therefore can take place before peak strength is reached. Figure 2.15b highlights another aspect of clay stress-strain behaviour, namely, its response to different strain rates. Two curves are shown, one for a triaxial test carried out in five minutes on a soft clay and another conducted over ten hours on the same material. The five-minute test clearly demonstrates that it is possible to increase applied stress well beyond the long-term yield stress, having the effect of increasing the rate of yielding. Strength and failure become much less easy concepts to grasp in clays, since they are both dependent on strain rate. For example, significant yielding occurs before peak strength is reached and it becomes difficult to say at what point failure occurred. Similarly, Fig. 2.15b shows that peak strength is strongly related to strain-rate and it is difficult to decide what may be truly representative of the material.

The property of yield at stresses lower than that necessary for failure in clays is usually attributed to a degree of viscous behaviour in the material. Indeed, clays can undergo considerable strains at low stresses without any sign of failure in the plastic sense and this phenomenon is called *creep*. Creep is a slow, continuous, viscous strain under low stress and is analogous to flow in a fluid. Above a certain level of applied stress, the upper yield point, clays tend to deform essentially plastically, and the yield strength corresponds roughly to that at which shear failure occurs. Below the upper yield point, applied stress causes creep at a rate which is dependent on stress (Fig. 2.15c). This is a truly viscous deformation, with strain continuing as

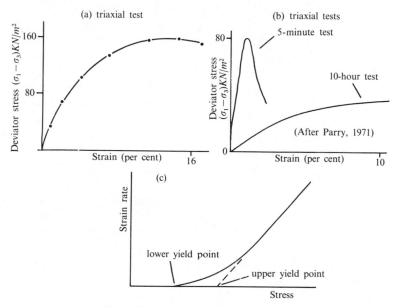

FIG. 2.15. Behaviour of clay under stress.

long as stress is applied, and so the strength concept has little meaning for the material. There may be a lower yield point, below which no creep occurs, though in some cases this may be very close to zero. Relatively weak electrochemical bonding between clay particles which are continuously breaking and reforming are probably responsible for creep behaviour.

Finally, it should be said that many attempts have been made to design theoretical models of sediment behaviour under stress. In order to accommodate the variability of sediment, viscous, plastic and elastic components have been combined into rather complex mechanical models. More recently, it has been recognized that variations in water content of a sediment also influence behaviour under stress and attempts have been made to incorporate these into general stress-strain models. An example is referred to in 'Further Reading' below.

Further Reading

CARSON, M. A. (1971a). *The Mechanics of Erosion*. Pion, London.

GRIFFITHS, J. C. (1967). *Scientific Method and the Analysis of Sediments*, McGraw-Hill, New York.

KONAMURA, F. and HUANG, R. J. (1974). 'A New Rheological Model for Soil Behaviour', *Proceedings of the American Society of Civil Engineers*, GT7, Paper 10675, pp. 807–24.

SMITH, G. N. (1974). *Elements of Soil Mechanics for Civil and Mining Engineers* (3rd edition), Crosby Lockwood Staples, London.

WHALLEY, W. B. (1976). *Properties of Materials and Geomorphological Explanation*, Oxford University Press, Oxford.

3 Transfer through Soils

This chapter is concerned with the movement of dissolved material and sediment particles through the pore-space system of soil. Although material in transport through soil pore-spaces is almost entirely derived from within, either by solution or by suspension of particles in the pore water, we shall be regarding the soil mass itself as a static filter. The fact that the entire mass of the soil might itself be in motion is not an issue for this chapter, but is taken up later, especially in chapter 5. Although the soil particles themselves are considered static with respect to movement through the pores, they are not regarded as inert. Interactions, both chemical and physical, between the immobile soil mass and substances in the pore-spaces are of paramount importance in determining their ability to move.

Except in very mountainous districts, most sediment must pass into a soil-system before entering a mass movement or fluid transport system. Direct erosion from bedrock is of small importance and so the processes of weathering and soil-formation are those which determine the quality of material eventually transported from an area. In the case of solutional transport, soil is of even greater importance since practically all solutes are derived from within the soil profile or just beneath it. The definition of soil used here is a little wider than that acceptable to most pedologists, since factors which have an influence on sediment and solute transport are most important to the discussion. In the case of movement through pore-spaces the factors most deserving attention are those which determine, and those which are dependent upon, pore-space geometry, such as porosity, permeability and grainsize distribution. A marked change in all these factors occurs at the interface between bedrock and overlying unconsolidated sediment. For the most part, in this chapter, it is assumed that the overlying material is *in situ* weathered regolith. That is, soil is considered to be a residual deposit, derived from the bedrock and released by processes of subaerial weathering. A pedologist would limit a definition of soil to the thickness of regolith which has been affected by downward acting processes to produce distinct layers or horizons, recognizable on chemical or physical variations with depth. In many cases the whole of a regolith has been so affected, and the definition of a soil used here is synonymous with that used by soil scientists. The distinction between regolith and bedrock is usually fairly sharp, not a gradual transition, and this is probably due to a very marked increase in bedrock weathering-susceptibility as soon as small cracks and joints are opened in the intact rock. It is fair to point out that the sediment veneer is not always derived from the underlying rock but may have been transferred from some point upslope and redeposited at a new site. Although no specific reference is made to such materials, much of the discussion applies equally to *in situ* and transported deposits.

A soil develops from bedrock through the action of weathering processes. Frequently lower plants, such as lichens, aid early soil development by colonizing rock surfaces. They increase chemical attack on the rock and provide a certain amount of organic material for other plants, possibly mosses, to colonize. They in turn produce more humus and increase chemical attack of the bedrock by providing organic acids as well as by lengthening the time that water remains in contact with the rock surface. The processes of weathering, residue accumulation and successive plant colonization continue until a regolith develops. (This is a very brief and simplistic résumé of an exceedingly complex interrelationship between biological and weathering processes.)

Soil-forming processes are made up of a number of vertical and lateral, or downslope, movements. Rainwater falling on the soil is divided into a lateral surface-flow component, capable of transferring material from the soil/air interface, and a vertical downward flow through the soil which picks up and carries material internally. Surface flow velocities are much greater than subsurface ones and therefore *overland flow* is in contact with the soil for a shorter time. It therefore has much less opportunity to dissolve material. Downward movement of material in percolating rainwater leads to losses in the upper soil where chemical equilibrium may be approached, and to possible net gains by reprecipitation in the lower layers if chemical environment changes with depth. These transfers result in vertical changes in colour, grainsize and chemistry and give soil its characteristic layered or *horizoned* appearance. In arid areas, where evaporation is high, the dominant direction of water movement through soil is upwards and causes enrichment in the opposite direction to that in humid-climate soils. When percolating water reaches the weathering front it is diverted laterally due to the abrupt decline in permeability, carrying with it solutes and perhaps some sediment derived from within the soil and from the weathering front. Eventually the lateral soil/water flow discharges from the soil system into a stream, with its load of suspended and dissolved material. In addition to vertical and lateral movements in water through soils there are a number of direct transfers of soil grains not dependent on water flow. Animals, for example, move large amounts of sediment towards the surface, tending to disrupt soil horizons. There are vertical heaves due to freeze–thaw or to moisture content changes, and lateral shifts of the whole soil mass in soil creep and landslides (chapters 4 and 5). The soil layer is therefore a very complex and active system with many inputs and outputs of water and material (Fig. 3.1).

Soil accumulating by weathering cannot continue to increase in depth indefinitely. Eventually outputs from the system must balance inputs from fresh bedrock weathering and from external deposition, should it occur. When a balance is struck, the regolith is in *equilibrium*, and it is generally true that equilibrium will always be approached, if not actually attained. A

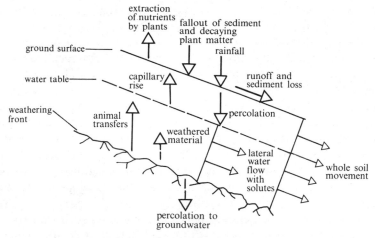

Fig. 3.1. The soil system.

couple of hypothetical examples will serve to demonstrate this point. First, consider a soil whose depth is steadily increasing. As time passes, percolating water will have progressively more time to attain chemical equilibrium before reaching the weathering front. At some depth, there will be no more dissolving power left and so bedrock weathering will cease until soil depth declines. The reverse argument is also true for a declining soil depth. Hence, soil depth will adjust to an equilibrium value where the inputs and outputs are balanced. Secondly, consider the case of a regolith being rapidly removed by transport processes. If weathering rate cannot increase sufficiently to balance this removal, all the soil will be stripped and will leave a bedrock surface exposed. A new equilibrium is established at zero soil depth. Here are two basic sediment derivation surfaces: the soil system where weathering can balance removal, and the bedrock surface where potential removal rate exceeds maximum weathering rate.

3.1. Solute transport through soils

The soil factors which reduce surface runoff of rainwater are generally those which favour solutional loss from soils, due to the increased importance of subsurface flow. Vegetation cover increases the permeability of surface soil horizons and reduces the velocity of surface flow, and both factors tend to increase the amount of water percolating into the soil (section 6.3). Since subsurface flow is many orders of magnitude slower than surface flow it is in contact with soil grains for longer and has more time to reach chemical equilibrium with soluble components. For example, consider a 100-metre-long slope at an angle of $20°$, with an effective soil grainsize (D_{10}) of 0·5 mm, an extremely coarse grainsize. Despite the coarseness of this soil, fully saturated flow would take about two days to reach the slope base. This is generally sufficient time for appreciable solution to take place, and one can assume for all practical purposes that subsurface flow is near-saturated with available solutes by the time it enters a stream. Overland flow would probably take less than three hours to flow over the same slope, depending on surface roughness and vegetation cover. Further, on a well-vegetated slope the flow would be insulated from direct contact with the soil and would have little opportunity to pick up material.

The more soluble bedrock components are removed directly at the weathering front and do not pass into the soil at all. Less soluble substances do persist in the soil but are slowly removed in time. Consequently, soils should show increasing concentrations of the least soluble materials through time, and with increasing rates of water flow (*leaching*) through them. The ratio of the proportions of a chemical in soil and in the original bedrock shows its persistence at any given time. This ratio is low for soluble materials which show little tendency to persist in soils, and high for relatively insoluble constituents. Through time, the ratio tends to decline exponentially for all components of the original bedrock, but at differing rates. Generally speaking, metallic cations like Ca^{++}, Na^+, Mg^{++} and K^+ are lost very quickly, silica (SiO_2) tends to persist for longer, while aluminium and iron hydroxides may be regarded as effectively insoluble. Hence, soils should shift in composition steadily towards silica and aluminium and iron hydroxides, and eventually to aluminium and iron hydroxides alone. The rate at which this is achieved is proportional to the degree of leaching in the soil, itself dependent on water flow. The trend is also to some extent dependent on original rock composition, shown by Table 3.1, which gives chemical analyses of granite and basalt and for soils developing from them. In both cases silica, aluminium and iron oxides make up the bulk of the derived soil (present in combined form in clay minerals) but loss of silica is greatest from the basalt. This is due to the fact that much silica in granite is present as almost insoluble quartz, whereas in basalt it is all combined in silicate minerals which are more easily weathered. It must be pointed out that 'solubility' is used in this section to imply the ease with which components are released into solution and does not imply a chemical process.

Another way of looking at the loss of chemicals from bedrock during soil formation is to compare the concentration of solutes in streams draining an area with their fractional concentration in the underlying bedrock. The ratio between these two quantities is called *mobility* and is usually expressed for oxides. It is not a constant for a given oxide but varies with the way it is combined and also with its concentration in the original rock. Average values are given in Table 3.2 for the oxides which make up over 95 per cent of dissolved metallic ions in rivers. A high mobility is indicative of low persistence in soils and so again soil composition should move progressively towards the bottom of Table 3.2 as time goes on. Verification of this trend towards high iron and aluminium hydroxide percentages through time is found in many humid tropical soils, where rates of *leaching* (solution by downward percolating water) are high. In this environment accumulations

Table 3.1

Analyses of granite and basalt and weathering products developing from them (by weight)

Oxide	Granite %	Clay derived[a] from Granite %	Basalt %	Soil derived[b] from Basalt %
SiO$_2$	73·6	67·5	49·0	10·9
Al$_2$O$_3$	13·8	26·6	13·7	28·8
Fe$_2$O$_3$	0·2 }	2·9	13·2	42·3
FeO	1·5 }		—	—
MgO	0·5	0·2	13·5	0·4
CaO	0·7	0	7·3	0·4
Na$_2$O	2·9	0·3	1·6	—
K$_2$O	5·0	1·5	0·3	0·3
TiO$_2$	1·6	0·5	1·7	8·1
Others	—	0·6	—	8·8

[a] Derived clay is 52% by weight of original rock from which it was derived.
[b] Derived soil is 23% by weight of original rock from which it was derived.

Sources: After data in Lovering (1958); Hatch, Wells and Wells (1961); and Buckman and Brady (1950).

of almost pure iron hydroxide soils (laterites) and aluminium hydroxide soils (bauxites) are relatively widespread and show the endpoint of a long period of intense leaching through soils. In temperate climates there has not usually been sufficient time for this endpoint to be reached at the slower rates of leaching and reaction which are obtained. This is especially so over much of northern Europe where the soils date from the last glaciation and are only a few thousand years old.

Components dissolved in soil water are carried in the direction of flow at the same rate, unless conditions change to cause redeposition. Downward solute movement or leaching in percolating water results in diminishing percentages of soluble components towards the surface, especially the most mobile ones, while lateral flow along the weathering front carries these out

Table 3.2

Mobilities of oxides based on average rock compositions and dissolved loads of rivers

CaO	3	Highly mobile
Na$_2$O	2·4	
MgO	1·3	
K$_2$O	1·25	
SiO$_2$	0·2	
Fe$_2$O$_3$	0·04	
Al$_2$O$_3$	0·02	Relatively immobile

Source: After Polynov (1937).

of the profile. The downward and lateral path of water and solute movement predominates wherever rainfall is in excess of evaporation and naturally leaching is most important in high rainfall conditions and highly permeable soils. Where evaporation exceeds rainfall there is a predominant, unsaturated, capillary flow *away* from the water table. Solutes are therefore carried upwards to accumulate near the surface and in contrast the more mobile ones are those most likely to accumulate. The mobility of specific ions which comprise the majority of the rock-forming silicate minerals is discussed below, with reference to experimental work and to chemical weathering processes.

The mobility of calcium, magnesium and sodium oxides is due to a number of factors. First, many minerals such as rock salt and gypsum are directly soluble in water. Calcium carbonate (calcite) and magnesium carbonate are not particularly soluble in pure water but are so in the presence of carbon dioxide, which is available in high concentrations in soil air due to micro-faunal respiration (section 1.6). However, these minerals form only small percentages of igneous rocks and so the high mobility of Ca^{++}, Na^+, Mg^{++}, K^+ must be due to another process. In this case it is due to the relatively easy substitution of hydrogen ions for metal ions in silicate minerals, the first stage of hydrolysis in silicate minerals such as felspar. Substitution is enhanced in acid conditions (section 1.6), which are supplied by organic acids in humus. Kerpen and Scharpenseel (1967) found that the mobility of Ca^{++} and Mg^{++} in laboratory leaching of basalt fragments was increased by more than ten times and that of Na^+ and K^+ by up to twice by adding a humus layer over the basalt. Under arid or semi-arid conditions even these mobile solutes can only be moved limited distances because the depth of water penetration is restricted by evaporation. For example, Jenny (1941) found that calcium carbonate accumulation in wind-blown soils in the U.S.A. increased in depth from only 25 cm with 35 cm annual rainfall to between 2 and 3 metres where rainfall was 100 cm. Where shallow groundwater occurs in high evaporation areas, upwards movement of capillary water causes concentrations of minerals towards the surface. Calcium carbonate, sodium chloride and calcium sulphate are among the more common surface mineral accumulations, but silica is by no means unknown.

The mobility of silica, though relatively small, is still appreciable and is impossible to explain in terms of the solubility of crystalline silica, quartz. It is probable that quartz saturation solubility is no greater than about 6 parts per million (ppm) by weight. Miller (1961) found concentrations of 4 ppm in streams draining a pure quartzite bedrock, a figure which agrees tolerably well with the theoretical maximum of 6 ppm. In rivers over the whole of the United States, however, silica averages around 15 ppm and can be higher (Davis, 1964), and experimental leaching of soil columns in the laboratory has yielded concentrations of over 20 ppm of silica in the drainage water. Clearly, these levels cannot be attributed to direct quartz solution and must be due to the solution of silicate minerals by the hydrolysis reaction, which releases soluble silica (section 1.6). Once again, leaching experiments have shown that overlying humus increases silica mobilization by enhancing hydrogen ion substitution. Experiments on soil pastes performed by Gifford and Frugoli (1964), showed silica to be in solution in soils as silicic acid (H_4SiO_4) which has a saturation concentration of between 23 and 28 ppm. This value is in close agreement to that obtained by leaching experiments on soils, and is not dissimilar to the maximum values found in streams.

The mobilities of iron oxide (Fe_2O_3) and aluminium oxide (Al_2O_3) are so low that sometimes they are regarded as standard constituents against which loss of other components may be assessed during soil formation (Ruxton, 1968). Paradoxically, although very small quantities of these oxides get into streams as dissolved load, they show appreciable mobility within the soil profile itself. This is especially true of iron oxide, which is commonly leached from upper soil horizons and transferred in percolating water to lower horizons where it is redeposited. Iron transfer is responsible for the pallid grey, iron-deficient, upper horizon and rust-red, iron-rich, lower horizon which is characteristic of many upland leached soils.

Iron and aluminium mobility within the profile has been explained in a number of ways. One suggestion is that iron is transferred in the reduced (ferrous, Fe^{++}) state, and indeed Fe^{++} is certainly very soluble over a wide range of pH. A problem which renders this explanation untenable in many situations is that ferrous iron is rapidly oxidized in aerobic conditions. Since many freely-drained upland soils show considerable iron translocation, another explanation must be sought. Aluminium (Al^{+++}) too, is largely insoluble under natural soil conditions. Recent work has shown that organic acid radicals are able to form soluble complexes with iron and aluminium and this is probably the way in which much transfer takes place. The reaction may be summarized:

$$M(OH)_n + Z^{-m} \rightleftharpoons MZ^{n-m} + nOH^-$$

iron or aluminium hydroxide	organic acid radical	soluble metal/acid complex	hydroxyl ions

A source of these organic acids, many of which are extremely complex, lies in the decaying leaf and humus layer which rests on top of the soil profile. A large number of laboratory experiments on leaching have now shown that extracts from many types of plant-debris contain organic acids which are capable of mobilizing iron in soils. The question arises: why is transfer restricted to the soil profile, and why do high concentrations of iron not appear in streams? The answer to this question is, of course, that chemical environment changes down the soil profile to favour the reverse reaction or else to destroy the organic radical. Several possibilities have been demonstrated. Results from leaching experiments indicate that drying out and allowing the soil profile to become aerated favour precipitation of iron hydroxide. The suggestion is that the reverse reaction is favoured in strong oxidizing conditions, but this reaction may be enhanced by micro-organisms attacking and breaking down the organic radical. It also appears that some of the complexes are extremely sensitive to pH changes and require only very small fluctuations for the complex to break down and iron to pre-cipitate. Finally, it has been suggested that some metal-acid complexes be-come unstable if the ratio of metal ions to acid radicals exceeds about 3 : 1, leading to precipitation. This is an attractive proposition, since it does not call for any *external* change in chemical environment but is entirely a pro-duct of the system. The organic acids are derived only at the ground surface, whereas iron and aluminium are available through the whole profile. Consequently, because there is a practically unlimited supply of Al^{+++} and Fe^{+++} ions the critical ratio is bound to be exceeded at some depth as the acid percolates downward and the mechanism is self-generating.

To summarize this short discussion on the transfer of solutes from soils, the following conditions are necessary for ion mobilization and removal:

(1) An appreciable downward and lateral flow of water is essential (in other words, a leaching environment). If there is little flow through the soil, water movement tends to be upwards due to capillary rise, and solutes accumulate instead of being removed.

(2) Chemical mechanisms by which ions can be mobilized are necessary. These are mainly the weathering reactions discussed in 1.6, of which hydrolysis is probably the most important because it is responsible for metal ion (Ca^+, Mg^+, Na^+, K^+) mobilization and silica (H_4SiO_4) solution. Hydrolysis is aided by an acidic environment, due to the presence of dissolved carbon dioxide or possibly organic acids. Organic complexing is important in mobilizing some ions, notably iron.

(3) There should be no tendency for conditions in the lower profile to favour redeposition, as in the case of iron and aluminium which are extremely immobile except under special circumstances. It is fair to point out here that redeposition of many ions does take place within the soil profile or downslope due to changes in chemical environment, but this is most important with iron and aluminium where re-deposition is practically complete.

3.2. Movement of sediment particles through soils

Clearly, the deciding criterion in subsurface sediment transport is whether par-ticles smaller than the general pore-space size are available for movement. In a well-sorted sediment, where the range of particle sizes is small, the pore-space will all be smaller than the smallest particles, and movement will be prohibited. A continuous distribution of sizes will also effectively restrict movement because each grainsize will be limited by the pore-space between particles slightly bigger than itself. Transport is most favoured if a bimodal size distribution exists, where the smaller particles can move through the pore structure of the larger ones. On geometric grounds alone, the size difference between soil-structural particles and those moving through the pores must be considerable. With perfect spheres the ratio of large : small diameter needs to be 2·5 : 1 for the loosest possible regular packing and 7 : 1 for the tightest, in order that the pores may be large enough for the small particles to fit through. Empirically, it has been found that the ratio needs to be even higher than these minimum values for free gravitational movement. One estimate (Statham 1974) is that particles do not move freely until the ratio is 15 : 1 for spheres in the dry state. Percolating water should facilitate the transport of particles through the pore-space of a soil, though the roughness of natural particles may in turn tend to inhibit motion.

The common occurrence of middle-soil horizons relatively enriched with clay sized sediment has, however, led many soil scientists to the conclusion that downward transfer of fine particles, if not lateral transfer, does take place through soil pores. High clay content, particularly the fine clay-fraction, is a consistent feature of the middle horizon of soils where downward percolation and leaching is intense, and so it is reasonable to equate enrichment with downward movement. Clay coatings are often observed on root channels and fissure surfaces and the fact that particles are oriented parallel to the channel with the coatings is indicative of downward transport. It should not be considered that the evidence for clay translocation through soils is incontrovertible. Brewer (1968) has estimated that only between 0–20 per cent of the increase in clay in middle horizons of some soils can be explained by downward movement. Differential weathering between the surface and lower horizons may be the main cause, or possibly external deposition of sediment on top of the soil in some cases. Even the presence of oriented clay coatings on fissure surfaces cannot be taken as cast-iron evidence of clay movement, since it appears they can develop in response to stresses imposed by expansion and contraction of the soil in wetting and drying cycles. Substantial downward movement has been noted in seasonally flooded soils in East Pakistan (Brammer, 1973). Here, coatings of clay and coarser particles develop extremely rapidly in from one to two years in soils subject to flooding. Mineralogy of the coatings is comparable with the topsoil, and Brammer suggests that particles are forced downwards under the high hydraulic head of the draining flood-water.

That downward movement of clay does occur to a greater or lesser extent is not really in dispute. From our point of view, we wish to know how fast it can occur and whether it takes place through primary pore-space or whether secondary pores such as fissures are required for it to be a viable sediment-transport route. These questions can be answered in part by the results of many laboratory experiments on the leaching of artificial soil columns. An interesting study was carried out recently on the movement of radioactively tagged barium sulphate crystals through a coarse loam (Kovenya, 1972). Particles larger than 10μ were almost immobile under most rates of flow through the soil but, strangely, 0.5μ particles moved faster than 0.1μ. This was attributed to physico-chemical absorption of the smallest particles on to soil grains. By far the most important factor in determining particle displacement was, however, the rate of water flow through the soil. Generally, particle transport rate was proportional to the rate of water flow but only about one-hundredth of its magnitude, indicating that particle movement is far from being an unimpeded suspension in the flow. The implication is that, for saturated percolation through a soil, downward particle movement rate would be 2 orders of magnitude slower than the soil infiltration rate. On a 100 metre-long slope, even with continuous saturated lateral flow, small particles might take up to 200 days to traverse the slope, whereas in surface flow they could take as little as two hours. On this comparison, subsurface sediment transport through *primary* pores must be of small overall importance.

Hallsworth (1963) conducted an exhaustive series of experiments on clay migration through sandy soils and investigated soil texture and chemical effects on movement. Movement declined rapidly with percentage of added clay and increased with the sand-size making up the soil skeleton. Hallsworth's data are summarized in Table 3.3 and it is apparent that the movement of montmorillonite is in all cases less than that of kaolinite, prob-

Table 3.3

The effect of clay percentage and soil matrix size on fine particle movement

Modal sand size (mm)	% weight loss (kaolinite)			% weight loss (montmorillonite)	
	5% clay	10% clay	20% clay	5% clay	10% clay
0.24	9	4	27	13	13
0.385	18	9	4	19	13
0.55	30	26	19	18	14

Source: After leaching experiments by Hallsworth (1963).

ably due to the capacity of montmorillonite to swell appreciably on wetting. Hallsworth also studied the effect of chemical environment on clay movement but found no simple relationships. Goss *et al.* (1973) have shown that surface-fissuring is of overriding importance in transmitting clay particles, rather than the primary pore-space. They found that rates of added clay movement through natural soils became negligible if the structure of the upper horizon was destroyed by mixing.

On two counts then, sediment movement through the primary pore-spaces of a soil can be considered unimportant in the transport of sediment down slopes. First, velocities are very low indeed. As demonstrated above, velocity of sub-surface particle movement may be only 1/2500th of surface suspended-particle velocities over a similar slope. Also the carrying capacity and velocity of subsurface solute load is much greater than subsurface sediment load. Secondly, the theory of filters predicts that the percentage of holes remaining unblocked in a filter declines exponentially through time. Thus, downward clay movement is a self-restricting mechanism with declining numbers of filter channels through time. Pore-spaces will get smaller and rates of water movement will decline with declining permeability if appreciable clay movement occurs, thereby restricting further movement through the soil. But if of limited importance in transfer processes, the mechanism is significant in soil profile development for the very reason that it can produce a lower permeability horizon at depth. Once again, we have the same situation as for dissolved iron, where a soil component may be mobile within the soil profile and be important in horizon-forming, but much less important in transport of material on a wider scale.

Experimental evidence supports the field observations on clay coatings that flow down cracks and fissures is much more significant in downward sediment movement than flow between grains. In semi-arid areas where long, dry periods are interspersed with extremely intense rainfall, sediment movement down fissures and along subsurface flow-lines sometimes becomes a dominant route for sediment removal from slopes. Ruxton (1958), for example, noted appreciable subsurface washing out of weathered material from a granite regolith in the Sudan. At times subsurface flow may be powerful enough to erode tunnels or *pipes* in regolith or soft sediment which can sometimes be large enough for human access. Many instances of piping of this type have been noted, and there seems to be general agreement that drying out of the soil and subsequent cracking play a major role in determining vertical flow routes. Waer is forced to flow laterally at some depth and causes subsurface erosion of fine material along pre-existing secondary permeability lines. Undoubtedly, the presence of highly soluble components in the sediment aids the development of secondary permeability but this does not seem to be a necessary prerequisite. Animal burrows may also provide lateral flow lines, but these too need not necessarily be present. Drew (1972) draws an analogy between pipes and cave systems in limestones. They have similar features, such as well defined 'sinks' or closed depressions where flow enters the ground; they form integrated conduit networks; and passage morphology is comparable. Rhoades and Sinacori's (1941) explanation for limestone cave development seems appropriate here, except that erosion is by sediment removal and not solution. Water sinks at well-defined points and flows laterally along the local water table to an outlet. The pipe itself is then thought to develop from the outlet back up slope by progressive erosion by the outflowing water. Flow in pipes is certainly turbulent rather than laminar (chapter 6) since the depths and velocities are comparatively great. It is probably better to consider them as rather special extensions of the stream network along secondary permeability channels into the soil, rather than as part of the soil-flow system as such, since they have much in common with stream channels.

This section has mostly been about downward and lateral transfers of sediment in soil water. But sediment movements do take place through the soil in an upwards direction, mainly under the influence of animal activity, and this can be locally important in overriding the visible effects of downward translocation. Space does not permit a lengthy section on *bioturbation*, as animal disturbance of soil is called, but a couple of examples will serve to show how it operates. Certain species of earthworm in Britain, and especially those which live in lowland soils, ingest soil material at depth but eject their casts at the surface. The rate of renewal of topsoil by this process may be as much as 0·63 to 6·3 cm depth in ten years (Atkinson, 1957). This is vastly in excess of soil horizon-forming processes and may be expected to obliterate horizon development, though worms are much less common and active in the well-drained acid soils of upland areas which shown the most marked horizonization. Whether worms play much part in direct bedrock weathering is rather doubtful, and their activities are probably confined to reworking the soil only. Termites in Australia have been observed to deposit about 2 to 3 cm of new topsoil to the surface every 1000 years by building their mounds, which are slowly broken down through time (Williams, 1968). Although this is a much lower rate of topsoil renewal than by worms in Britain, it is more significant in sediment derivation, because similar termites in Nigeria have been shown to transport up to one-third of the new topsoil directly from the weathering front as weathered bedrock fragments (Nye, 1955). Thus termites not only stir up the soil profile; in the Nigerian case they also add about 2·5 cm of new soil per 1000 years and are responsible for accelerating the rate of bedrock lowering by transferring partially-weathered material away from the weathering front. It may be that other animals also transfer partially-weathered bedrock to the surface. Soil animals also play a part in soil creep, which is discussed in chapter 5.

3.3. Conclusion

Soil plays a very important part in the transition from weathering to transport in most geomorphic systems. Rates of movement of constituents out of the soil by all transport processes are such that they remove soil no faster than it is produced by bedrock weathering. As already mentioned, if the rate of removal is too fast the soil system is destroyed, but since bare rock slopes are a small part of most landscapes it can be assumed that potential rates of transport from soil systems do not often exceed maximum rates of bedrock weathering.

Everything which aids subsurface water-flow increases solutional transport at the expense of sediment transport from the soil. High soil permeabilities increase the percentage of rainwater which travels through the soil rather than overland. Subsurface velocities are low and there is very good contact between sediment and the water, so one would expect soil water to approach chemical equilibrium with whatever soluble components are present. Surface flow is rapid and has rather poor contact with sediment grains and may therefore be appreciably undersaturated with available solutes. Quite the reverse is true of sediment transport. Surface flow rates are rapid and therefore able to pick up and carry sediment easily. Subsurface flow is constrained to move slowly and can only entrain very fine clay particles, and there are also restrictions imposed by the size of pore-space through which grains must pass. Subsurface flow is likely to supply a very small percentage of the total sediment load of streams on account of its low particle-travel velocity and the extremely small size of particles transported. Only in conditions where secondary permeability lines, such as pipes, develop is subsurface sediment yield likely to be appreciable, and in these circumstances it seems more reasonable to consider the subsurface flow to be a part of the stream network.

Further Reading

BIRKELAND, P. W. (1974). *Pedology, Weathering and Geomorphological Research*, Oxford University Press, New York.

BLOOMFIELD, C. (1951). 'Experiments on the Mechanism of Gley Formation', *Journal of Soil Science*, **2**.

——(1953, 1954, 1955). 'A Study of Podzolisation', *Journal of Soil Science*, **4**, **5** and **6**.

CARSON, M. A. and KIRKBY, M. J. (1972). *Hillslope Form and Process*, chapter 9 'Subsurface Water Erosion', pp. 231–71, Cambridge University Press, London.

GOUDIE, A. (1973). *Duricrusts in Tropical and Subtropical Landscapes*, Clarendon Press, Oxford.

STOBBE, P. C. and WRIGHT, J. R. (1959). 'Modern Concepts of the Genesis of Podzols', *Proceedings of the Soil Science Society of America*, **23**, 161–4.

4 Transfer of Sediment by Rapid Mass Movement

A relatively consistent distinction exists in sediment transport processes between those in which large quantities of sediment move together in close grain-to-grain contact and those where transport is mainly in the form of single grains dispersed through a fluid transporting medium. The former are often called *mass movement processes* and are discussed in this chapter and in chapter 5. The distinction between mass movement and flow processes is to some extent genetic, and is one which is extremely convenient for discussion purposes. But it is important to remember that, on close examination, sharp distinctions between the two broad process-groups merge into a gradual transition (Fig. 4.1). The continuum between flow and mass move-

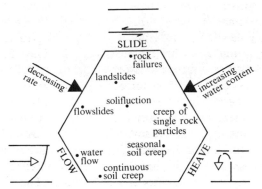

FIG. 4.1. A classification of mass movement processes. (After Carson and Kirkby, 1972)

ment processes makes a purely rational treatment of transfer processes rather difficult and some genetic relationships have been sacrificed to expedience in this book. In particular, flowslides, some of which can justifiably be argued to be dispersed flow processes, are dealt with in this chapter for reasons discussed later. Also the process of rockfall (section 4.4) is treated as a mass movement simply because it is difficult to classify in any other way.

In this chapter we are dealing with rapid mass movements, which are generally distinguished by the occurrence of rapid failure when a finite yield strength in the material is exceeded. Slow mass movements or creep processes (chapter 5) involve no such critical yield strength in the medium. Again, this is a convenient rather than an entirely consistent distinction and often it is impossible clearly to separate these types of behaviour in some materials.

4.1. Styles of rapid mass movement and their analysis

A useful concept, the *factor of safety concept*, can be employed in the analysis of rapid mass movements, provided a specific yield strength is applicable to the failure in question. Factor of safety is defined as the ratio of shear strength to shear stress on a potential shear plane:

$$F = \text{shear strength/shear stress}$$

and a value of unity is an indication of a potentially unstable condition. Engineers use the concept to assess the stability of designed slopes, and also apply it to natural slopes when the need arises. We shall now consider major styles of landslide, and how safety factor is evaluated for them.

Landslides are mass movements within a soil or rock where shear is confined, more or less, to a well-defined shear or slide plane. They are usually separable into shallow or deep-seated failures according to whether cohesion is present as a measurable strength factor at failure or not. As a general rule, cohesionless media fail as shallow slides with a ratio of length/depth to shear plane of 10 to 100, whereas cohesive sediments suffer deep-seated failures with the ratio nearer to 2 to 5. This is, of course, an extremely broad generalization to which many exceptions occur, but it is a useful first approximation. Slope angle is the only geometric control on slope stability in cohesionless materials. If the angle is greater than a critical value the slope is unstable throughout its depth. Thus failure will pass through the weakest horizon near to the ground surface. A cohesive material, unlike a cohesionless one, possesses some strength at the ground surface, but if the rate of shear-stress increase with depth exceeds the rate of shear-strength increase, there will be a *critical depth* below which it becomes unstable (Fig. 4.2). Thus, in a cohesive soil, height of slope as well as slope angle is a controlling factor, and deep-seated failures are more common.

Dry, cohesionless soils rely only upon their interparticle frictional strength for stability. Since stability is independent of soil depth or slope

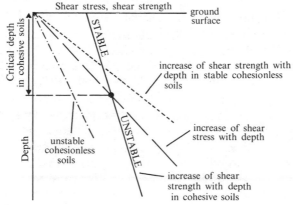

FIG. 4.2. Increase of shear strength and shear stress with depth beneath a slope.

height, failures are usually shallow and roughly parallel to the ground surface. Although shallow slides are never perfectly planar slips of a soil-mass
over a smooth, flat, shear plane they can often be approximated as such. The
forces existing in a so-called *infinite planar slide* are shown in Fig. 4.3 and it
is easily shown that the stable slope angle is equal to φ (angle of internal
shearing resistance). The analysis in Fig. 4.3 is quite adequate for many
cohesionless slides, even though it is an oversimplification of reality.

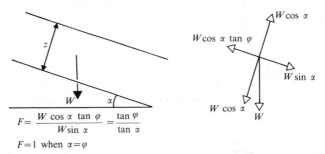

$$F = \frac{W \cos \alpha \tan \varphi}{W \sin \alpha} = \frac{\tan \varphi}{\tan \alpha}$$

$F = 1$ when $\alpha = \varphi$

FIG. 4.3. An infinite planar slide in cohesionless soil.

The geometry of a slip in homogeneous cohesive soil is strongly related to
slope form, but two representative simple forms are considered here. In a
vertical bank (Fig. 4.4), stress directions are roughly analogous to a triaxial
test (chapter 2) where the shear plane must be inclined at $(45 + \varphi/2)°$ to the
principal plane. Principal stress (σ_1) acts vertically through the bank, and so
the shear plane ought to be approximately straight at an angle of $(45
+ \varphi/2)°$ to the horizontal. These assumptions are often used to calculate the
critical height of a vertical cliff in a method of analysis called the *Culmann
Stability Analysis*. Strictly speaking, the Culmann approach is only applicable to vertical slopes, where principal stress operates vertically throughout,
and where the shear plane is flat.

When slope angle is less than vertical, the simple stress system illustrated
in Fig. 4.4 for a vertical bank no longer applies. Principal stress direction
(σ_1) tends to bend towards the slope surface until it acts parallel to the slope
at the ground surface itself. Again assuming that the failure plane must
always be inclined at $(45 + \varphi/2)°$ to the principal plane, which is valid if the
medium is isotropic, the shear plane will be curved (Fig. 4.5a). Failures of this
type are quite characteristic of cohesive media and are called *rotational slips*.

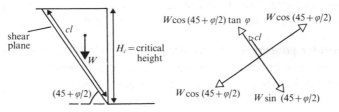

FIG. 4.4. The stability of a vertical bank.

Rotational slips can be analysed in a number of ways, the simplest of which assumes pure rotation about the slip plane centre O, the so-called *circular-arc* method, in which moments are taken about O in the horizontal and vertical direction (Fig. 4.5b). More recent methods have tended to subdivide the sliding mass into a series of slices, analyse the forces acting on each slice and then sum these over the whole slip (Fig. 4.5c). The method of

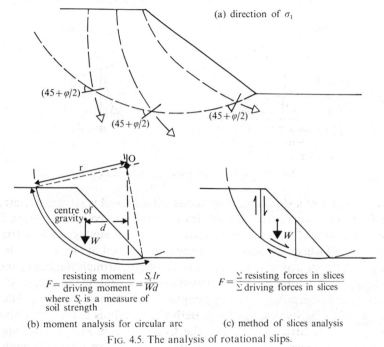

(a) direction of σ_1

$(45+\varphi/2)$

$(45+\varphi/2)$

$(45+\varphi/2)$

$$F=\frac{\text{resisting moment}}{\text{driving moment}}=\frac{S_c lr}{Wd}$$
where S_c is a measure of soil strength

(b) moment analysis for circular arc

$$F=\frac{\Sigma \text{ resisting forces in slices}}{\Sigma \text{ driving forces in slices}}$$

(c) method of slices analysis

FIG. 4.5. The analysis of rotational slips.

slices is more popularly used for slopes where pore-water pressures need to be considered (section 4.2).

Before continuing to describe the influence of pore–pressures on slope stability, it seems only fair to point out that deep-seated failures are by no means as simple as the preceding section may imply. Whenever a line of weakness, structural or stratigraphic, exists in a sediment, failure is likely to be deflected along that line. Thus there may frequently be considerable deviation from a straight or circular arc failure plane. The models discussed above have, however, been found to be reasonable approximations to actual conditions existing in real slides in a wide range of different materials.

4.2. Pore-water pressure and shear strength—the effective stress concept

Hitherto we have only hinted at the role played by pore-water pressure in shear strength, but its role is of paramount importance in landslides. Pore-water pressure is, fairly obviously, the pressure exerted by the pore fluid within the pore-space system. Pore–pressure is negative when the sediment is unsaturated; that is to say a suction is exerted on the pore-space. Conversely, saturated pore-space is subjected to positive pore–pressure. The effect of pore–pressure on strength is basically rather simple. Pressure is exerted on the soil grains, tending either to force them apart or to draw them together, and therefore increases or reduces the normal stress between them. Thus the *effective normal stress* (σ') becomes:

$$\sigma' = (\sigma - u)$$

where u is the pore pressure. Coulomb's failure law becomes:

$$s = c' + \sigma' \tan \varphi'$$

and the dashes denote that pore–pressures have been taken into consideration. In unsaturated conditions negative pore–pressures increase the normal stress and therefore the strength, but this condition is not usually of interest to us because positive pore-water pressures usually occur at some time in the history of a slope to reduce the total strength. It is the weakest link, or most critical pore–pressure, which actually decides whether a failure will take place or not. Pore–pressures can be extremely variable according to topography and anisotropy of the soil and so we shall look only at some simple pore-water distributions which have been found to approximate in many different slides.

Drained Pore–Pressure Conditions

Drained pore–pressures exist when the pore-water is able to flow freely through the pore system without building up excess pore–pressures. Thus, pore–pressure is never greater than the observable head of water acting at any point. On a fully submerged slope, both the normal stress *and* the shear strength are proportionally reduced by buoyant upthrust, and stable slope angle is the same as for a dry slope.

It is common for a marked decrease in permeability to occur at a roughly constant distance below the ground surface, corresponding to the soil/bedrock interface. This is especially true of weathered regolith overlying coherent bedrock and it results in downward-percolating rainwater being forced to flow laterally, parallel to the permeability break. Consequently, lateral *throughflow* in a saturated zone at the base of the soil is a common groundwater flow condition on many slopes. Consider a soil mantle of unit weight γ (unit weight is weight/unit volume) in which groundwater is flowing (Fig. 4.6a). The head of water at depth z is indicated by the level of water in a standpipe inserted to that depth and is equal to:

$$u = \text{pore–pressure} = \gamma_w z_w g \cdot \cos \alpha \qquad (4.1)$$

where γ_w is the unit weight of water.

Hence effective normal stress $= (\sigma - u) = (\gamma_s g z \cdot \cos \alpha - \gamma_w z_w g \cdot \cos \alpha)$
$$= (\gamma_s z - \gamma_w z_w)g \cos \alpha. \qquad (4.2)$$

Shear stress along the failure plane is not reduced because no component of the pore pressure acts parallel to it.

$$\text{Shear stress} = \gamma_s gz \sin \alpha \qquad (4.3)$$

$$\text{and factor of safety} = F = \frac{(\gamma_s gz - \gamma_w z_w) \tan \varphi'}{\gamma_s z} \frac{}{\tan \alpha}. \qquad (4.4)$$

The worst possible or critical groundwater condition is when the water table is at the surface. Although this may be rare on some slopes, it needs to happen only once to be the cause of limiting stability. At this point, $z_w = z$ and equation (4.4) simplifies to:

$$F = \left(\frac{\gamma_s - \gamma_w}{\gamma_s}\right) \frac{\tan \varphi}{\tan \alpha} \quad \text{or} \quad \left(1 - \frac{\gamma_w}{\gamma_s}\right) \frac{\tan \varphi'}{\tan \alpha} \qquad (4.5)$$

and

$$\tan \alpha_{\text{crit}} = \left(1 - \frac{\gamma_w}{\gamma_s}\right) \tan \varphi' \qquad (4.6)$$

Since density of soil is around 1·9–2·1 and that of water is 1, γ_w/γ_s is about 0·48–0·52. Critical stable-slope gradient for a cohesionless slide with parallel groundwater flow is, therefore, close to $\frac{1}{2}\tan \varphi$. Actual cases where this condition has been found to prevail are considered later in the chapter.

It is much more difficult to generalize groundwater flow conditions in sediments where no permeability break occurs parallel to the ground surface, and yet this information is vital to a stability analysis of a rotational slip. However, the water surface is usually convex upwards, grading to a base level at the slope foot (Fig. 4.6b). If the groundwater surface can be

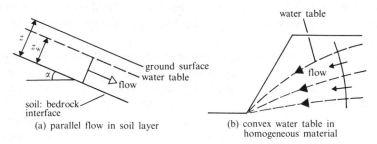

(a) parallel flow in soil layer   (b) convex water table in homogeneous material

FIG. 4.6. Groundwater flow in soils.

measured, using a series of standpipes, it is possible to compute the effective normal stress for each slice of a rotational slip, for example. Frequently, it is not feasible to measure the water table, for example in the case of a slide which was related to a water table which no longer exists, and it is then necessary to resort to assumptions about the water surface.

Higher pore-water pressures can develop in circumstances where ground-water flow is upwards through a sediment, resulting in a water table *above* ground level (*artesian conditions*). Although this may not be very common, it may be locally important in slope stability where water is confined within a specific layer. Freezing of the ground surface has been noted as a cause of artesian conditions in the lower part of the soil profile.

Undrained Pore-water Conditions

Although undrained pore-water conditions are not particularly common, they do occur at times and have a pronounced effect on soil strength. A soil may be regarded as *undrained* if pressures imposed on the pore fluid cannot be rapidly dissipated by flow through the pore-space system. It is a condition restricted mostly, but not entirely, to fine-grained, clay-size sediment where permeability is low. By implication, undrained conditions may only exist on the short time-scale because even the finest materials are to some extent permeable and are able to dissipate excess pore pressure with time.

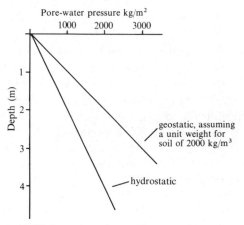

FIG. 4.7. Hydrostatic and geostatic pore-water pressures.

Having said that, Vaughan and Walbancke (1973) have shown that where a load is taken rapidly off unweathered London Clay (by removal of overburden in a cutting, for example) it may take between thirty and a hundred years for the negative pore-water pressures induced to be dissipated and an equilibrium value reached. Stresses seldom change so rapidly in natural sediments, and so undrained conditions are infrequent. Undrained conditions develop in fine-grained soils because pore-water cannot drain rapidly out of the medium when loaded. Since the soil grains are infinitely more compressible than water (which is effectively incompressible), the increased pressure is passed directly to the pore fluid.

To take the example of a water table in a horizontal bed of sediment, normally pore-water pressure at any depth z is equal to the head of water above it (Fig. 4.7). This is the *hydrostatic* state in which pore–pressure is equal to $\gamma_w z$. In the worst possible undrained condition no water can escape from the soil, and so when load is applied *all* the weight of the soil is borne by the pore-water fluid. In this case, the *geostatic condition*, pore–pressure is

$\gamma_s z$, and will be about twice hydrostatic, assuming a unit weight for saturated soil of 2000 kg/m³. With this pore–pressure distribution, equation (4.6) becomes:

$$\tan \alpha_{\text{crit}} = \left(1 - \frac{\gamma_s}{\gamma_s}\right) \tan \varphi' = 0 \qquad (4.7)$$

for perfect geostatic conditions.

4.3. Instabilities in clays

A very important factor in the strength of clay sediments is the degree of overburden pressure to which they have been subjected. A *normally consolidated* clay has never been subject to greater overburden pressure than exists at the present. In other words, the present ground surface has never been any higher and no loads due to external forces have been applied to it. This state corresponds to the degree of consolidation at time of deposition. In contrast, an *overconsolidated clay* has at some time in the past experienced greater loads than it bears at the present. A former peak load may have been due to overburden pressure of overlying strata, since eroded away, or it may have been the result of external influences. Many Mesozoic and Tertiary clays in Britain, for example, have been subjected to very great overburden pressures which are of course now entirely missing along their outcrops. Glacial till deposited under an ice sheet (*lodgement till*, section 6.6) has been overconsolidated by the external pressures of ice once above it and by forces exerted on it by ice movement. Commonly these forces are complex and the degree of overconsolidation in a till may not bear any simple relationship to the former depth of ice. Most clays are to some extent overconsolidated; if they were not their surface layers would be little more than 'clay soup'. The degree can vary, however, from negligible amounts to an extremely large part of the peak strength.

Overconsolidation causes a decrease in porosity, expulsion of pore-water and a concomitant increase in particle interlocking and interaction. Consequently, cohesion (c') and internal shearing resistance (φ') can be expected to increase. The effects are illustrated graphically in Fig. 4.8.

As Fig. 4.8 suggests, shear strength of a normally consolidated clay increases linearly with depth, and cohesion is more or less zero. Overconsolidated clays, on the other hand, possess cohesion which is directly relatable to the overburden pressure removed. In the simplest case of vertical erosion:

$$c' = \gamma_s z_e$$

where z_e is the depth of overburden removed and γ_s is the unit weight of the soil. In practice, it is probably never feasible to relate cohesion quite so simply to overburden even if z_e were known. The water-content curve, too, is truncated in an overconsolidated clay, giving rise to much lower water content than in normally consolidated clay at the same depth.

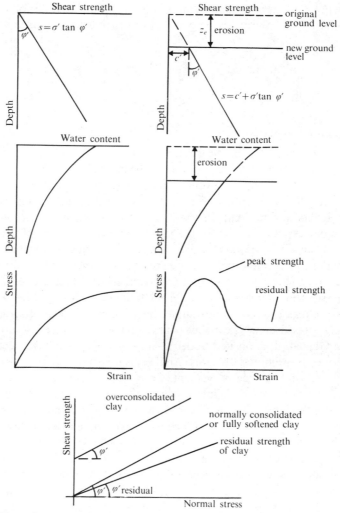

FIG. 4.8. Normally consolidated and overconsolidated clay strength. (After Skempton, 1964)

Quite different stress-strain behaviour is apparent in overconsolidated clay. A marked peak in the stress-strain curve of an overconsolidated clay is followed by a rapid fall after failure, to residual levels at high strains. The fall is equatable to loss of cohesion and a fall in frictional strength from φ' to φ'_{res} (residual angle of internal shearing resistance) along the shear plane.

From Fig. 4.8 one may expect normally consolidated clays to be essentially similar to any other cohesionless materials; that is to say, their strength is related only to friction and pore-water pressures at failure. Consequently, shallow landslides are the dominant style of failure, the stable angle being modified by pore pressures. Overconsolidated clays present a much less simple behavioural pattern and the precise style of slip depends upon what strength parameters actually operate at failure. This is expanded in the following paragraphs.

From the discussion of landslide styles in section 4.1, it would seem that deep-seated failures of the rotational type should be the major style of landslide occurring in overconsolidated clays, on account of their cohesive strength. Indeed, failures of this kind undeniably occur and three examples are listed in Table 4.1. Of these three examples, two were in boulder clay (Selset, Glencullen), one in Norwegian marine clay (Lodalen), and all were in unfissured materials. That is to say, there were no large-scale cracks or joints passing through them. Also, all occurred on natural river bluffs, though the Norwegian example had been modified by man. From Table 4.1 it is evident that for these slips, which were *first-time slips in undisturbed sediments*, the factor of safety calculated from peak strength parameters (c', φ') was close to unity whereas it was much less than unity when residual strength was assumed (c'_{res}, φ'_{res}). These slips were related to the *measured peak strength* as one would expect, and they were rotational slips. After the first failure, shear strength is lost along the shear plane and ultimately falls to residual levels in

Table 4.1

Six failures in overconsolidated clays

Location	Material	c' (kN/m²)	φ'	c'_{res}	φ'_{res}	F(intact)	F(residual)	Style	Reference
Selset (England)	Boulder Clay	8	32°	—	30°	~1·0	0·69	Rotational	Skempton and Brown (1961)
Lodalen (Norway)	Marine Clay	10	27°	—	—	~1·0	—	Rotational	Sevaldson (1956)
Glencullen (Ireland)	Boulder Clay	variable	30–35°	0	28–32°	~1·0	0·6	Rotational	Statham (1975)
Jackfield (England)	Coal Measures	7	25°	0	19°	1·45	1·07	Shallow	Henkel and Skempton (1955)
Uppingham (England)	Lias Clay	17	23°	0	18°	—	0·93–1·02	Shallow	Chandler (1970)
U.S.A.	Clay Shales	variable	20°	0	9–17°	1·4–1·6	1·0–1·1	Shallow	Noble (1973)

the slipped debris. Failure style in intact, overconsolidated clay media follows a pattern to be expected from their strength parameters: first-time slips are deep-seated and related to peak strength; subsequent slips demonstrate diminishing cohesion and a fall to residual strength parameters, establishing shallow slips as the dominant style.

Such a pattern of failure in overconsolidated clay is not always the case, especially if *fissures* are present. A series of puzzling *first-time* slips in overconsolidated London Clay and other strata was noted by Skempton (1948) in which the stability of artificial cuttings was time-dependent. Steeper cuttings were stable only for a short time but shallower-angled slopes could fail after up to a hundred years of stability. In all cases the strength at failure was much less than the measured peak strength. Natural slopes, too, show failures on slopes that should be stable if peak strength had operated

(Henkel and Skempton, 1955; Skempton and Delory, 1957) and, further-more, the failures are *shallow parallel slips* and not deep-seated failures. One of these, the Jackfield slide (Henkel and Skempton 1955), provided an op-portunity for detailed analysis of the conditions at failure. The Jackfield slide was a shallow failure in a fissured, overconsolidated, Coal Measure clay-shale, whose shear plane was more or less planar and parallel to the ground surface. Strength parameters for the clay are listed in Table 4.1. It is ap-parent that failure occurred at a level of strength approaching residual strength, for a *first-time failure*. The peak strength for the material was there-fore reduced *without a visible failure occurring*. Henkel and Skempton (1955) concluded that fissures allowed water to percolate into the soil, causing softening and weathering along the fissure lines to such a degree that co-hesion was effectively reduced to zero and a shallow slide ensued. Subsequently, many workers have found that the ultimate stability of many fissured, overconsolidated clays is controlled by residual and not peak strength. Two examples are quoted in Table 4.2.

Table 4.2
Strength parameters for two overconsolidated clays

	Intact strength		Remoulded strength		Fully softened strength		Ultimate stable slope
	φ'	$c'(kN/m^2)$	φ'_{res}	c'_{res} (kN/m^2)	φ'	$c'(kN/m^2)$	
London Clay (Skempton and Delory, 1957)	25°	12	19°	0	25°	0	9–10°
Lias Clay (Chandler, 1974)	23°	17	18°	0	23°	0	8–9°

Recently, it has been suggested (Skempton, 1970) that first-time slips in overconsolidated clays are controlled by the fully softened strength rather than the remoulded strength. The fully softened strength is reached at the point where further strain produces no further increase in water content. This is the critical pre-failure state, roughly corresponding to zero cohesion and peak φ', and it is attained after very small strains. Full residual levels are not reached until much larger strains have occurred. The process of slow loss of strength through time is called *progressive failure* and is due to in-creasing water content through time as softening takes place. Fissures play an important role in the softening process by acting as concentrators for stress at certain points and by allowing weathering processes and perco-lation water to penetrate the sediment.

Fissures may have a more immediate effect on clay strength if they are so well developed in the unweathered state as to constitute significant planes of weakness. When this occurs, cohesion may be effectively zero across the fissures even when no softening has taken place, with important implications for stability and failure style. Esu (1966) found very well-pronounced, almost orthogonal, joint sets in some lake and marine clays in Italy, resulting not only in short-term stability but also in a departure from the more normal rotational slide-type. Cuttings in lake clays failed soon after excavation as shallow slab slides along the discontinuities. Pore-pressures within the fissures aided instability, a feature common in harder rocks but less so in clays. Marine clays studied showed toppling failure (page 88) owing to softening and expansion along joint lines. The difference between these two types of failure indicates the importance of fissure orientation with respect to slope orientation. Fissure orientation in glacial tills is related to the direction of ice movement, with joint sets often forming normal to the direction of ice travel. They are therefore due to overburden pressure imposed by a moving ice sheet, and can sometimes be responsible for slopes being preferentially less stable at certain aspects.

In the case of fissured, overconsolidated, clay slopes subjected to rapid basal erosion, weathering processes may never have the opportunity to soften the clay before a new surface is exposed by slipping. The most obvious examples of rapidly retreating clay slopes are coastal, such as are commonly found in clay cliffs of southern England. In these cases, stability is always controlled by intact strength and once again, rotational slips prevail.

4.4. Landslides in weathering regoliths

One type of weathering regolith, that accumulating on overconsolidated clay, has already been the subject of discussion in section 4.3. Weathering softens the clay along fissure lines, reducing strength to the fully softened condition, and subsequent failure reduces the strength further to residual levels. Failures in regoliths from harder rocks can also be influenced by weathering through mechanisms rather different from those acting in clays.

Regolith consists of a residual accumulation of minerals and rock fragments remaining after weathering, with an increasing percentage of bedrock fragments as the weathering front is approached. Since it consists of an assemblage of loose particles released from bedrock it is an almost cohesionless material in which one would expect shallow slides. Shallow slides generally strip the entire regolith from the bedrock since the weakest horizon is very close to the weathering front. This is possibly for two reasons: first, the upper soil layers are bound by plant roots; and secondly the permeability break (section 4.2) at the weathering front concentrates any lateral flow at the base of the regolith. Consequently, sliding sequences in regolith consist of long periods of weathering separated by short periods of slipping in which a new bedrock surface is exposed. The rate of removal by these slips is thus closely controlled by the rate of weathering. Shallow slips occur on regolith in all climatic environments, but are especially important in the humid tropics.

Slope angles at which shallow sliding takes place are listed for a number of localities in Table 4.3. It is clear from the data that stability is closely controlled by the pore-pressure conditions which are obtained in the most critical situation. For example, where pore pressures are low or absent, maximum stable slope angle is high and approaches φ'. Slope gradient is about $\frac{1}{2}\tan\varphi'$ where parallel groundwater flow to the ground surface occurs (section 4.2) but can be much lower where high (for example artesian) pore-water pressures are generated. Even in the cases where pore-water pressures are not important to stability they can be important trigger mechanisms, and so there is a strong association between failure and periods of heavy rain.

Table 4.3

Shallow failures in regoliths

Locality	Unstable angle	φ'	φ_i	Pore pressure condition at failure	Surface lowering rate
Hawaii (Wentworth, 1943)	42–48°	—	—	? small	10 cm/130 yrs
New Guinea (Simonett, 1967)	35–36°	—	—	? small	10 cm/440 yrs (non-tectonic) 10 cm/70 yrs (tectonic)
Ireland (Statham, 1975)	35–38°	34–37°	—	? small	
Pennines/ Exmoor (Carson & Petley, 1970)	25–27°	40–44°	—	parallel flow to ground surface	—
Laramie Mts. (Carson, 1971)	23°	—	44°	parallel flow to ground surface	
Spitzbergen (Chandler, 1972a)	8–12°	36°	—	Artesian (rising pore-water flow)	—

Changes in shear strength due to weathering

The changes brought about by weathering in clays can be illustrated with reference to two papers by R. J. Chandler (Chandler, 1969; 1972b) in which the properties of Lias Clay and Keuper Marl in different weathering states are analysed. Chandler differentiated four zones of decreasing weathering from the surface, ranging through completely weathered to fresh clay which was encountered at anything up to 100 metres depth. Significantly, there was no tendency for the weathering zones to run parallel to cutting-surfaces up to 120 years old, showing that the properties which serve to distinguish the zones play no direct role in the progressive failure mechanism (at least in the medium time-scale). Notable changes with weathering include:

(1) increasing oxidation—determined visually by the presence of iron staining and analytically from the $FeO : Fe_2O_3$ ratio.
(2) increasing clay-sized particle content—10–35 per cent in unweathered clay, rising to 30–50 per cent in fully weathered clay. The increase in clay was due to the breakdown of particle aggregates rather than individual particles.
(3) increasing water content—water content rose from 15 per cent in the unweathered state to 40 per cent when fully weathered.
(4) increasing numbers of discontinuities and shear planes towards the surface.

The Lias Clay and Keuper Marl are both heavily overconsolidated clays and some of the changes noted above play a large part in the softening process discussed in the previous section. In particular, the changes in water content and clay content appear to be important in this respect. Chandler found for the Keuper that cohesion declined from over 28 kN/m^2 to less than 17 kN/m^2 and peak φ' from more than 40° to 32–35°, as weathering progressed. More significantly, φ'_{res} declined from 23–32° to 18–24° through weathering. This is indicative of a decline in strength of *normally consolidated* marl through time, quite independent of the softening process leading to progressive failure. This may be due to increasing clay content through time (Fig. 4.9).

Mechanical changes due to weathering in regoliths forming on harder rocks are no less dramatic. As Fig. 2.10 shows, experimental work verifies that sediments of mixed grainsize (coarse/fine size mixtures) have a peak strength corresponding to a coarse fabric with its pore-spaces completely filled with fine material (taluvium). *Discontinuous breakdown* of sediment,

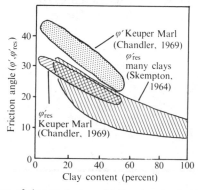

FIG. 4.9. The influence of clay content on frictional strength. (After Chandler, 1969)

that is, mechanical disintegration of coarse particles into fine grainsize particles, will give a progression from coarse gravelly material through taluvium to a sandy or silty sediment. When *continuous breakdown* occurs, coarse particles are as likely to split in half as to produce fines, resulting in a continuous size distribution rather than the bimodal distribution of discontinuous weathering. Consequently, pore-spaces will not become filled in the same way, and we may expect much less dramatic changes in sediment strength to ensue. Of course, many processes occurring in weathering are being neglected here and we are only concentrating on the aspect of changing grainsize distribution through time; that is, mechanical weathering. Chemical changes occurring in the regolith may also greatly influence strength. For example, clay minerals may be formed and effect a decline in frictional strength. Equally, the particles making up the matrix may become weakened by chemical attack to the extent that failure planes begin to pass through, rather than around, grains. This would lead to a reduction in particle-interlocking with weathering, an effect which has been observed by Nishida and Kagawa (1972) in some residual soils. With these points in mind, however, one may still expect to find that grainsize distribution-changes through weathering have an influence on shallow landsliding.

Critical weathering states in soils: the threshold concept

With continuous weathering, fine material is produced through time, and mean grainsize declines, but there is no opportunity for the pore structure to become blocked with fines. Changes in sediment properties are therefore steady. Porosity declines a little; permeability and shear strength decline slowly. In the case of discontinuous weathering there is a high probability that small particles will be produced by weathering and a concomitant low probability of grains splitting in two. Hence the size distribution becomes progressively bimodal, the macropores fill with fines, and the progression from gravel to taluvium will be accompanied by a rapid decrease in porosity and permeability and an increase in frictional strength.

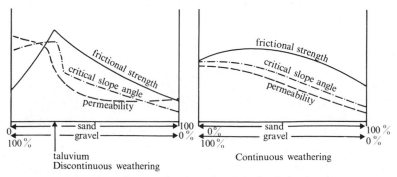

FIG. 4.10. Changing soil properties through weathering in hard rock regoliths.

What relevance might these changes have in the initiation of shallow landslides? With continuous weathering, the decline in frictional strength and in permeability will progressively increase the risk of high pore-water pressures with high rainfall, and a more or less steady decline in stable slope angle is to be anticipated through time. Consequently, events should only be small, continual adjustments to the slowly changing soil parameters. In contrast, discontinuous weathering will lead to critical weathering states, through which soil properties rapidly decline, followed by periods of little change. Just such a critical state corresponds to taluvium, whose increase in frictional strength is more than compensated for by a massive decline in permeability, increasing the pore-pressure risk. Slips may therefore be discontinuous, larger events which are associated with the critical weathered state (Fig. 4.10).

Carson and Petley (1970) and Carson (1971) have studied the distribution of slope angles in areas of weathered regolith on different bedrocks. In all areas there was some tendency towards grouping of slopes at particular angles (Table 4.4). The 33–34° group corresponded to accumulated coarse

Table 4.4

Commonly occurring straight slope angles in some weathering mantles

Location	Lithology	Common slope angle modes				
Pennines	Gritstone	20°	*GAP IN DISTRIBUTON*	27°	*GAP IN DISTRIBUTION*	33°
Laramie Mts.	Granite/Granite-gneiss	20°		27°		33°
Exmoor	Shale	20°———→ 28°				34°

Sources: After Carson (1971b), Carson and Petley (1970).

regolith or scree. Slope angles between 32° and 29° were relatively uncommon but another mode occurred at 27° and again at 20° on granite and sandstone bedrocks. On shales, however, these last two modes fused to a continuous distribution from 28° to 20° in which all slope angles were represented. Carson and Petley (1970) proposed that the 27° and 20° groups were related to threshold conditions in taluvium and sandy regolith respectively (in both cases the critical stable slope gradient is about $\frac{1}{2}\tan\varphi$) for discontinuous weathering. Shales showed a continuous distribution of angles because they were more susceptible to continuous weathering and hence gradual changes in critical stable slope angle.

4.5 Instabilities in hard rocks

Rock strength factors

We shall take a somewhat broader definition of 'hard' rocks than the normally accepted geological definition and apply the term to any cemented or lithified rock. The important property of hard rocks is that individual grains are held together by extremely powerful mineral cements, or by having been fused together on cooling from a magma. Grains are so strongly held together that their contacts may be as strong as the grains themselves. Intergranular cohesion is so great that grain breakage occurs as readily as sliding along grain boundaries. Consequently, these internal boundaries are less important, grain interlocking is effectively reduced, and the material behaviour approaches that of a homogeneous continuum. The division into cohesive and frictional strength becomes much less easy to visualize in the failure of a hard intact rock, and often strength is quoted simply as the compressive strength (S_c) of a cylindrical rock specimen. Compressive strengths in hard rocks can be extremely high (Table 4.5).

Using the following expression for stable cliff height (H_c):

$$H_c = S_c/\gamma \quad \text{(Terzaghi, 1962)},$$

where γ is the unit weight of the rock, very large stable heights of thousands of metres are obtained even for relatively weak rocks (Table 4.5). This formula gives a worst possible case where a tension crack separates the potential rock failure from the cliff face. Rock slopes are nearly always unstable at heights only a fraction of their stable height when calculated from intact compressive strength. Hence one may question the applicability of this parameter to the strength of large rock masses in the field.

Hard rocks practically never exist without structural breaks or discontinuities such as joints, bedding planes and faults. These provide pre-existing planes along which cohesive strength is considerably reduced and which facilitate sliding. They are internal boundaries in much the same way as grain-to-grain contacts in granular media, except that they are on a larger scale, across which cohesion and friction once again can be appreciated as separate elements of strength. Joint-cohesion is, however, only a small fraction of intact rock compressive strength, a factor which reduces the maximum stable cliff height in hard rocks dramatically (Table 4.5).

In most rocks, discontinuities are not distributed isotropically, but have a pronounced preferred orientation. Bedding planes form continuous parallel breaks at roughly uniform spacing, and joints develop in well-defined sets with a preferred orientation and dip, often bearing a simple geometric relationship to fold-structures in the rock. Joint orientation with respect to a cliff has a marked influence on the style and scale of rock failures, often aided by pore-water pressures within the joint system.

Table 4.5

Representative strength parameters for some rocks

Rock type	Intact rock			Rock with Discontinuities		
	Unit Weight gm/cm^3	*Compressive Strength* S_c (kg/cm^2)	*Critical Height of Vertical Cliff* H_c (metres)	*Cohesion* kg/cm^2 c'	φ *Degrees*	[a] *Critical Height* (metres)
Granite	2·614	1000–2500	4000–10 000	1–3	30–50	12–65
Sandstone	1·950	200–1700	1000–9000	0·5–1·5	30–45	9–40
Shale	2·400	100–1000	400–4000	0·2–1·0	27–45	4–20
Limestone	3·169	300–2500	900–8000	0·25–1·0	30–50	5–25
Quartzite	2·614	1500–3000	6000–11 000	1–3	30–50	12–65

[a] Based on Culmann method, assuming material behaves as isotropic continuum of discontinuities and assuming tension crack develops.

Classification of failures in hard rocks

Three factors are of recurrent importance in the classification of failures in hard rocks:

(1) presence of discontinuities
(2) extent to which discontinuities are weakened by weathering
(3) scale of the event.

Fig 4.11 is an attempt towards a genetic classification, partially based on these factors. A great number of terms have been applied to rock failures, but many are confusing and contradictory, and so only a few simple descriptive terms are used in this book. A comprehensive genetic classification is not really possible with the present, somewhat poor, state of knowledge concerning the mechanics of some rock-failures, especially the larger ones. For this reason Whalley (1974) has preferred to classify only on the basis of scale of event, although this system is not followed here.

Rockfall is a small-scale process operating on almost all rock walls in some degree, regardless of the over-all stability of the slope. It is a process by which relatively small blocks of rock are detached from the cliff face, falling freely through space to the foot of the cliff. Rockfall can take place only under certain conditions: small joints and fissures are essential and the rate of joint-opening must be substantially in excess of the rate of grain detachment. Without these conditions, loose blocks cannot be produced. The importance of freeze–thaw activity in loosening and detaching joint blocks has been indicated in a number of studies of rockfall frequency, notably by

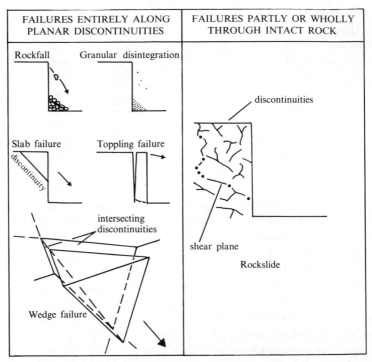

| FAILURES ENTIRELY ALONG PLANAR DISCONTINUITIES | FAILURES PARTLY OR WHOLLY THROUGH INTACT ROCK |

FIG. 4.11. Failure styles in hard rocks.

Gardner (1970) and Bjerrum and Jorstad (1968). Gardner, working in the Canadian Rockies, noted that 75 per cent of all rockfall events occurred on north-east-facing slopes, which experience lower temperatures since they face away from the sun. There was also a marked diurnal variation in frequency, with the peak corresponding to the rise in temperature just after midday, when thawing-out was complete. Expansion of the heated, loosened blocks may also have some part to play in their final release. High rockfall frequency has been reported by Bjerrum and Jorstad (1968) from Norway in the spring and autumn, when numbers of freeze–thaw cycles may be expected to be greatest. Rockfall fragments fall through space and accumulate at the cliff-base as *scree* or debris slopes, unless they are removed by a transporting agent such as a glacier or river.

Toppling, slab and wedge failures are small- to moderate-sized failures and are characteristic of well-jointed and bedded rocks. The failures take place almost entirely along joint-lines, which may have been opened to a greater or lesser extent by weathering, and so style of failure is modified by joint orientation. *Toppling failure* is likely when consistent discontinuities are present or develop parallel to a vertical cliff face. The discontinuity may be a pre-existing joint or bedding plane in vertically inclined strata, which is progressively opened by weathering processes until the point of equilibrium is passed. Alternatively, jointing develops parallel to a cliff face due to lateral

relief of earth pressure, brought about by erosion. The development of
stress-relief joints or *tension cracks* is greatly aided by basal undercutting of
the potentially unstable face and this is achieved in a number of ways. A
large and well-documented toppling failure, the fall of Threatening Rock on
the Colorado Plateau, was advanced by undercutting in a shale-bed beneath
the sandstone monolith (Schumm and Chorley, 1964). Koons (1955) has also
noted that cliff collapse in New Mexico is accelerated by weathering of an
underlying, weak, shale horizon. Coastal sites probably furnish the fastest
rates of basal undercutting, aiding the rapid development of tension cracks.
Hutchinson (1968) has noted such a process in the chalk cliffs of Kent.

When bedding planes are inclined towards a face, there is an opportunity
for *slab failure* to take place. Slab failure is basically a very simple process in
which sliding takes place along the inclined bedding plane. Consequently,
stability is controlled by friction and cohesion along the bedding and pore-
water pressures in the discontinuity system. The factor of safety for a sliding
slab of weight W is:

$$F = c + \frac{(W \cdot \cos \alpha - u) \tan \varphi_\mu}{W \cdot \sin \alpha}$$

where c is cohesion, α is the bedding dip, u is pore-water pressure and φ_μ is
the angle of plane-sliding friction of the rock surfaces. Thus slab failure is
mechanically analogous to the simple sliding of a block on an inclined plane
(chapter 1). Once failure has occurred on a bedding plane, cohesion is of
course lost, and the angle of plane-sliding friction declines to a residual level
owing to wear on the rock surfaces.

Discontinuities frequently form almost orthogonal intersecting patterns
which cut the rock into blocks. When such a pattern of structural breaks is
inclined obliquely to the cliff face, the potential for *wedge failure* exists. The
geometry of a wedge failure is illustrated in Fig. 4.11 and it is clear from the
diagram that, for a failure to be possible, the mutual intersection of the
planes delineating the wedge must dip towards the face at a lower angle than
the face.

Rockslides include all those failures where at least part of the failure
plane passes through intact rock. The simplest failure one could imagine
would be that of a vertical cliff in perfectly homogeneous, unjointed rock.
As we have already mentioned, the intact strength of most hard rocks is
so great that critical cliff heights could never be attained. However,
Lohnes and Handy (1968) have studied the failure of vertical river banks in
loessic (wind-blown) material in Iowa and Tennessee and, while loess hardly
fits the description 'hard rock', the failures do at least illustrate the principle
involved. Essentially, Lohnes and Handy used the Culmann stability analysis (section 4.1) and with the measured soil parameters the predicted stable
bank height was in close agreement with the observed. Furthermore the
exposed shear plane, which became the river bluff on removal of the slipped
mass, was close to the expected inclination of $(45 + \varphi/2)°$. In the case of

harder, non-stratified rocks with some discontinuities, Terzaghi (1962) suggested that measured cohesion must be replaced with effective cohesion, which allowed for the percentage of the shear plane following pre-existing joints.

Occasional, but extremely large, rockslides do occur in which it is very difficult to analyse the mechanics of the conditions leading to failure. They are most common in deeply glaciated valleys where stress relief parallel to the valley sides may cause tension cracks along which failure may partly take place. Undoubtedly pore-water pressures in the joint systems may be very important, and these have been discussed at length by Terzaghi (1962). However, in a review of large-scale rockslides, Whalley (1974) points out that some workers found levels of stress in rocks to be much greater than their overburden pressures would suggest. These high levels of stress may be due to remnant tectonic stresses and may be highly significant in large rock failures.

4.6. Mobilization and flow of slide debris after failure

The vast majority of slope failures entail only relatively small movements along the shear plane before stability is regained and movement ceases. In clay materials, rotational slips may move as little as a few centimetres before sliding ceases, and it is rare for more than a few metres' movement to take place unless basal erosion continues to maintain an unstable slope. Cohesionless slides may travel further, but usually stop quickly when the slope becomes stable or when the slipped mass runs on to a lower slope angle below. Thus movement is confined to the limits of the unstable slope and rarely exceeds a matter of metres. In all cases, rates of movement are in terms of hours or days, rather than seconds or minutes.

On occasion, however, landslide debris becomes highly mobile after slipping, sometimes to the extent that it travels as a fluid. The reasons for fluidization will be discussed below but it is clear that travel distance and velocities can be very much greater if fluidization can be achieved. Strictly speaking, many of the processes discussed below are not mass movements, since the genetic criterion of grain-to-grain contact is not met. Many should be regarded as dense flow processes, but since they are most commonly associated with an initial landslide, and are composed of slipped debris in a mobile state, it seems more logical to regard them as an extension of the landslide process. Furthermore, highly mobile slipped debris is not always in a viscous state, and so to classify them all as such would be misleading.

Slow mudflows

After a rotational landslip has taken place in a clay material, stability is usually regained quickly and the rotated block retains much of its original form. In certain circumstances, usually where no vegetation protects the clay surface from direct weathering, slipped debris is broken up by raindrop erosion, slopewash processes and expansion/contraction forces. This is especially true of coastal cliffs where basal undercutting maintains a relatively steep angle. The slipped debris, and other debris falling from the cliffs, is steadily comminuted to a disorganized mass of hard clay-fragments in a matrix of softened clay whose water content is very high.

Rotational slipping on coastal cliffs results in the formation of bowl-shaped hollows, termed 'corries' (Hutchinson, 1970) by analogy with the glacial landform. The form tends to concentrate weathered clay debris in one spot, from whence it moves in a 'glacier-like' manner down towards the beach. Slips on the headwall supply small feeder-flows, whose angles are around 13°–18°, which in turn coalesce to form the thicker accumulation flow at an angle as low as 4°. Pore-water pressures in the accumulation flow bear no relation to those in the feeder flows, which tend to be related to general groundwater patterns in the cliff. The slow mudflow is therefore a steady debris-transfer system in equilibrium with the rate of supply from the cliff and removal at the base. They are stable at moderate rates of basal removal where almost all cliff retreat is indirect via toe-erosion of the flows (Hutchinson, 1967). At higher rates of removal, the mudflow equilibrium is destroyed and cliff retreat is by direct attack on the intact cliff.

It is now time to qualify the term 'slow' mudflow, used to distinguish these flows from the more rapid flows described below. Hutchinson (1967) re-corded an average velocity of 13·3 m/yr on a coastal mudflow in London Clay in Kent, equivalent to a discharge of 930 m³/yr of clay. Very much greater velocities up to 85 m/yr were noted over short periods. Prior and Stephens (1972) have recorded mudflow movement in Northern Ireland on a continuous basis, and the pattern revealed is very interesting. Their records show that flow velocities may be constant over long periods but, in detail, movement is made up of a series of almost instantaneous jerks and is not continuous. Movement is also rather seasonal, confined usually (but not exclusively) to the winter months. Hutchinson (1970) found that most move-ment in the flows was in fact, basal slip on a shear plane. These observations imply that movement is rather more akin to a shallow slide than to a flow.

What might be responsible for such low-angled slide behaviour in softened clay debris? Hutchinson and Bhandari (1971) quote two examples from Bouldnor Cliff in the Isle of Wight where soil and topographic para-meters are known (Table 4.6). These flows should be stable at about 7° if simple parallel groundwater flow conditions prevail, and yet their stable angles are around 4°. Hutchinson and Bhandari (1971) suggested that un-drained pore-water pressures induced by rapid addition of debris to the flow might be responsible for low stability. Indeed, pore-pressures recorded by them in a mudflow recently loaded by the addition of new debris were close to geostatic (section 4.2). Prior and Stephens (1972) and Prior *et al.* (1970) found mudflow advances were related to intense rainfall periods and it may be that addition of new debris occurs mainly in storms, or that high pore-pressure on the basal shear plane may be caused by heavy rain.

Table 4.6

Mudflow data from Bouldnor Cliff (Isle of Wight)

	Measured slope angle	φ'	c'	Stable slope assuming parallel pore-water flow to ground surface
Bouldnor 1	4·5°	13·5°	—	7°
Bouldnor 2	3·9°	13·2°	—	7°

Source: After Hutchinson and Bhandari (1971).

Quick clay flowslides

Quick clay, so called because of its tendency to liquefy on disturbance, is a common deposit in Southern Scandinavia and Canada but is also known in other parts of the world. Its ability to behave as a fluid is stunningly illustrated by the famous Vaerdalen Slide, which occurred near Trondheim in 1893 (Holmsen, 1953). The Vaerdalen slide involved over 55 million cubic metres of debris and left a crater 2·9 km² in area. The liquid clay spread over 8·5 km², ponded a sizeable lake up behind itself, and travelled a maximum distance of over 6 km from the slip zone at a speed faster 'than any horseman could ride'.

A quick clay slide usually begins as a straightforward rotational slip on a steep river bluff. After failure, however, the slipped mass tends to liquefy, flow from the slide zone and expose another steep face which in turn slips and liquefies. This process of rotational slipping, liquefaction and subsequent slipping of the next exposed face may continue for some distance and it is called *retrogressive failure*. Mobilization of the debris commonly occurs on slope angles of 3–6° but it has been noted on slopes as low as 1°, indicating that the clay may possess almost no strength at all after disturbance. The soil parameter which best describes the behaviour of quick clay is the ratio of undisturbed to remoulded shear strength, called sensitivity (Skempton and Northey, 1952). Remoulding breaks down the structure of a soil. Normally consolidated clays have a sensitivity around 1 to 2; for overconsolidated clays it is rarely more than 10; but quick clays have sensitivities of up to 100. The implications of a drop in shear strength on remoulding to one-hundredth of the original strength are all too apparent.

Quick clays are fairly uncommon on a world-wide basis and must be considered an unusual occurrence. Consequently, theories proposed to explain their behaviour will only be discussed briefly here. There are two main theories: the leaching theory (Rosenqvist, 1953) and the more recent mineralogic theory proposed by Cabrera and Smalley (1973). Rosenqvist argues that quick clays were deposited in a tranquil marine environment with a very delicate flocculated or 'card-house' arrangement of the clay particles. However, when raised above sea level, they remain stable due to the electrochemical bonding between the particles and saline pore-water, which is strongly electrolytic. Through time, the saline pore-water is leached out by rain water, reducing the interparticle bonding very significantly. Further, fresh water is free to circulate through the pore-space because it is less closely bonded to the grains, and so the effective water content increases. Consequently a slight disturbance, such as a rotational slip, causes collapse of the delicate structure and liquefaction. In support of Rosenqvist's ideas, considerable experimental work has shown that the strength of soil can be markedly increased by adding saline water and decreased again by leaching. Also, some indirect experimental evidence suggests that a very open grain structure exists in some quick clays.

Cabrera and Smalley's theory suggests that most quick clays contain a large percentage of clay-size quartz and felspar produced by glacial scouring. If so, there should be a predominance of inactive cohesive bonds (chapter 2.2) which are easily broken on slight disturbance. Thus, if a rotational slip occurs, the slipped mass will become a suspension of fine particles in the pore-water because the cohesive bonding will break. Once separated, the particles will stay in suspension for a long time on account of their small size. Again, some experimental results support Cabrera and Smalley's contention and no attempt is made here to suggest which of the two theories may be correct. It is possible that both are valid, but apply to quick clays in different areas.

Flowslides in coarse sediments

Coarse sediments show less tendency towards fluidization than fine ones because larger grainsizes cannot be supported in suspension for very long. Nevertheless flowslides do occur in coarse materials, though their variable characteristics make generalization difficult. One style of flowslide which has been noted in widely different climatic regimes and sediment types is the *debris flow*. Debris flows consist of fluidized lobes of poorly-sorted coarse and fine sediments which originate on debris-mantled hillslopes. The lobe moves at a few kilometres an hour, leaving low ridges of debris or levées (Sharp, 1942) on either side of its track as it goes. They range in size from one to ten metres across and carry from one to several cubic metres of debris. Indeed, the range may be even greater (100 000 m^3) if one accepts the Aberfan flowslide as being similar in nature. The track itself may or may not be eroded by the passage of a debris flow; indeed, it is common for a vegetation cover to be completely undamaged, which suggests that basal shear stress is not great. This in turn is evidence of viscous behaviour, with shear stress distributed through the flow rather than concentrated at the base as in a slide. Sometimes they occur as spatially and temporally discontinuous events, that is, the probability of a flow occurring on a previously undisturbed section of hillslope is as great as one occurring along a previous flow track (see Prior *et al.* 1970), but on other occasions they become concentrated along gullies (Winder, 1965; Statham, 1976). In the latter case, cones of debris are built at the gully mouth by deposition from the flows.

The potential of a material towards flowslide behaviour is indicated by its *brittleness index* (Bishop, 1973) which is:

$$I_b = \frac{\tau_f - \tau_r}{\tau_f} \times 100\%$$

in which τ_f is the resistance to shear at failure and τ_r is the residual resistance to shear. A large brittleness index is a sign that a large drop in strength occurs in the residual state. In sands, brittleness index increases with density of packing because low-porosity materials show a larger drop in frictional strength through failure. Debris which can be broken down by shearing may have high brittleness indices. Aberfan coal-tip debris, composed of coarse, shaly fragments before shearing, was reduced to a sticky, clayey mass along the shear plane during shear testing (Bishop, 1973). This change in state was accompanied by a decline in frictional strength from $\varphi' = 39 \cdot 5°$ to $\varphi'_{res} = 18°$, and the brittleness index was from 50 to 60 per cent.

Materials with high brittleness values are thus more susceptible to flowslide behaviour, but this is not the only factor causing this type of slide. For example, a flowslide in coarse-grained fluvioglacial sand and gravel is described by Statham (1975). Mean grainsize of the deposit was more than 1 mm, clay and silt content less than 12 per cent, and the brittleness index was no more than 15 per cent. In this case, the initial slide was caused by high pore-water pressures after snowmelt in a somewhat unusual topographic situation, and subsequent mobility was probably related to these high pore pressures being maintained by a constant supply of water from two small streams. The presence of spring lines was also an initiating factor in the Aberfan slide (Bishop, 1973) and in debris flows in North Antrim (Prior *et al.*, 1970). High rainfall has been an important trigger mechanism in many debris flows. For example, Prior *et al.* (1970) noted debris flows on days with over 37 mm of rain and also on days when hourly rates in excess of $6 \cdot 3$ mm/hr occurred. Again, such high rainfall intensities may cause pore-water pressures to build up, especially along pre-existing seepage or spring-lines. Generally, however, it must be emphasized that pore-pressures are only a trigger for the slide. Subsequent mobility depends on properties of the sediment itself.

Mobility is undoubtedly related to high water-content in the sediment. This may be derived directly from the high pore-pressures and water content in the debris prior to slipping, which appears to be the case with the North Antrim and Aberfan flows mentioned above. High water contents may also result from incorporation of water into stable debris, which causes a reduction in strength. Observations by Johnson and Rahn (1970) tend to suggest that debris slowly absorbs water from streams and rainwater until slipping begins, when the movement causes churning and more admixing of water to reduce the strength still further.

Generally, debris flows are initiated on fairly high-angled slopes, of the order of more than $20°$ for talus-like debris, and continue to move at constant velocity on slopes around 10°, depending on the material (Table 4.7). Possibly, motion continues as long as the slope is steep enough to maintain a velocity which precludes drainage of water from the debris. Once water is able to drain out, pore pressures subside and frictional strength is restored.

Table 4.7

Slope Angles on which flowslides are initiated and maintained

Material	Initial slope angle	Slope angle at which movement continued	Reference
Talus	25°	—	Prior (1970)
Talus	22°	16°	Statham (1976)
Coarse sand and gravel	—	11·5°	Statham (1975)
Sandy, clayey debris	21°	11°	Johnson and Rahn (1970)
Tipped mine waste (Aberfan)	—	12·5°	Bishop (1973)

This assertion is supported by the fact that debris flows drain and become firm if they are stopped by a blockage in their path.

Mobilization of rockslide debris

Infrequently, but often with disastrous consequences, the debris from large-scale rockslides becomes highly mobile after failure and can travel many kilometres, even on low slopes or upslope. Details of three such flowslides are listed in Table 4.8, but a description of just one of them will be useful in explaining their general characteristics, in so far as it is possible to generalize about a small number of highly variable events. The Mount Huascaran slide in Peru began as a rock failure triggered by an earthquake, high up on the summit face of Mt. Huascaran (Browning, 1973). It fell freely through space for 650 metres from its original altitude of around 6500 metres until it crashed into a ridge of glacial moraine, incorporating some of this material into the slide. It travelled the next 14·5 km on a 23° slope at the incredible speed of 400 km/hr and, although it demolished everything in its path, it was apparent from the lack of erosion of the ground surface that the slide was effectively frictionless at this point. It was while travelling at this high velocity that the mass demolished two towns and killed over 21 000 people. It continued to move for another 50 km at 25 km/hr on a slope of 5°, but by this time it was in contact with the ground surface and behaved as a water-saturated, dense, viscous flow.

Characteristics of the Mt. Huascaran slide and other slides of this nature can be summarized:

(1) They move at high velocity and have little frictional contact with the ground.

(2) There is little or no sorting of the debris and no abrasion of constituent blocks.

(3) When they come to rest after their period of rapid motion they form irregular, hummocky and low-angled debris piles, which often extend upslope.

Kent (1966) thought it possible that fluidization by entrapped air was responsible for mobilization of the debris, and quotes much eye-witness evidence from the Frank Slide in Canada to support the belief that air was included in the moving material. He suggested that the falling mass in-

Table 4.8

Size and mobility of three large rockslides

	Date	Volume $(m^3 \times 10^6)$	Distance travelled (km)	Mechanism	Estimated Velocity (km/hr)
Sherman Slide, Alaska (Shreve, 1966)	1964	30	5	air cushion	80–330
Frank Slide, Canada (Kent, 1966)	1903	35	4·5	air cushion	170
Mt. Huascaran, Peru (Browning, 1973)	1970	2	14·5 } 64·5 { (air cushion) 50 } { (fluid flow)		400 25

corporated air, which caused high pore pressures and reduced internal frictional strength. Shreve (1968) concurs with Kent that entrapped air is the cause of high mobility but doubts whether fluidization is due to high pore pressure. Shreve is of the opinion that a cushion of air is trapped at the base of the slide material and that, instead of fluidization causing the mobility, the debris moves as a plastic slide over a near-frictionless surface. Permeability of the debris must be low enough to ensure that air escapes only slowly from the cushion, otherwise rapid settling would take place. The air-cushion mechanism is a plausible one to explain the mobility of rockslide debris, but it must be admitted that the mechanics of the process are only poorly understood.

4.7. Controlling factors in rapid mass movements

Long-term and short-term trigger mechanisms

Rapid mass movements are caused by a change from stability to instability over a matter of hours or minutes. From the geomorphic point of view they can be regarded as instantaneous, though the factors leading to instability can themselves be changing gradually. There is then, almost always a recognizable *trigger mechanism* which is responsible for reducing factor of safety to less than one.

One of the points which should have become clear throughout this chapter is that all types of soils and rock at the earth's surface undergo changes in strength due to weathering. Overconsolidated clays progressively soften and lose cohesion, rocks weather along joint lines to lose their intact strength, and regoliths undergo many changes which affect their strength. These are long-term changes in strength which influence the rate of landsliding over periods of tens to hundreds of years, or even longer. Thus, the factor of safety for a slope may decline through time by weathering until the critical value is reached.

An engineer, however, is concerned with much more immediate causes of landsliding and is unlikely to look to weathering-related phenomena in order to explain them. He would generally regard soil parameters as being static, though an exception may be found in overconsolidated clays which show 'long-term' changes on a timescale short enough to be of interest to an engineer. The long-term changes can best be thought of as slow-burning fuses, rather than as trigger mechanisms. They steadily increase the susceptibility to failure but are never the recognizable cause of a slide. In practice almost all failures can be related to short-term oscillations in the safety factor caused by one of the following:

(1) Abnormally high rainfall intensities leading to high pore-water pressures.
(2) Seismic disturbance.
(3) Erosion at the base of the slope to increase its angle or height.

And of these the first is by far the most important. Anyone requiring specific examples of mass movements initiated by the above mechanisms need look only to some of the works quoted in this chapter, or to the UNESCO 'Annual Summary of Information on Natural Disasters' which has since 1971 included reports of landslides.

Long-term factors lead to a gradual increase in failure susceptibility, and decrease in factor of safety, through time which is diagrammatically illustrated in Fig. 4.12a. The time to failure is therefore a reflection of rate of change of soil properties to which rate of landsliding is tied. The role of the short-term rainfall factor is indicated by the dotted line in the diagram. It causes the safety factor to oscillate around the long-term change and, of course, failure will always occur at a low point in the oscillation when pore pressures are high. Therefore, failure will always be recognizably related to wet weather periods. 'Abnormally' high rainfall events usually occur often enough for there to be no scarcity of immediate trigger mechanisms. In fact, the use of the term 'abnormally high rainfall' is misleading; rather there are rainstorms with a given probability of recurring at some time interval. If one recalls that Prior et al. (1970) noted that debris flows occurred in Northern Antrim with hourly rainfall rates in excess of 6·3 mm/hr, and that this intensity occurred eleven times in the observation year, it is easy to see that trigger intensities are readily available. If a slope does not fail under such intensities it is probable that long-term changes have not reached the critical state. One can see then that the effective time to failure is advanced by pore-water pressures only by an amount equal to the recurrence interval of the trigger, which is usually not very long.

Seismic events are perhaps less predictable and reliable than high-intensity rainfall, even in areas of very high seismic activity. When they do occur, their importance in triggering mass movements is beyond question. In addition to the Mt. Huascaran slide in 1970, triggered by an earthquake off the coast of Peru, many thousands of small rockslides were triggered in the immediate neighbourhood. Seismic activity is also a consistent trigger in New Zealand and San Francisco Bay, among other regions. The fact that seismic disturbances are not so predictable and frequent as high rainfall, and that they can cause failure on slopes which were in themselves stable, makes them very interesting. The influence of seismic activity on factor of safety is shown in Fig. 4.12b. Disturbance can be sufficiently large to reduce time to failure very considerably, and rock surfaces are exposed to fresh weathering by sliding at a faster rate. As a first approximation, rate of increase of soil depth through time due to weathering declines to zero (Fig. 4.12c). This decline in rate is due to many factors such as protection of the weathering front by the regolith from direct weathering, and also the attainment of an equilibrium between increase in soil by weathering and decrease by erosion loss. If this is the case, seismic activity can increase the rate of landslide activity, by re-exposing rock surfaces to faster rates of weathering more

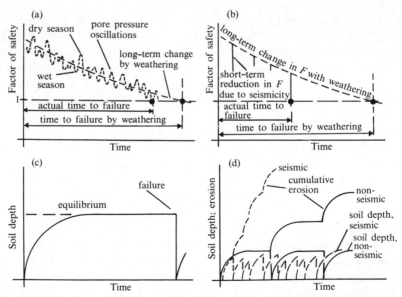

FIG. 4.12. Long-term and short-term trigger mechanisms for landslides. (a) changing factor of safety in a non-seismic region; (b) changing factor of safety in a seismic region; (c) increase of soil depth through time; (d) composition rules of erosion by landsliding in seismic and non-seismic regions.

often (Fig. 4.12d). There is little evidence for this process, because very few comparative studies have been made on rates of landsliding in comparable seismic and non-seismic areas. But there is one slim piece of evidence from New Guinea (Simonett, 1967), where rates of surface lowering by landslides have been estimated to be four to five times as high in seismic as in non-seismic areas (Table 4.3).

To conclude, we may think of long-term factors as being the ultimate control on landsliding rate, and short-term triggers the initiating factor only. Thus, a particular landslide-style is an equilibrium process when it removes the same amount of material from a slope over a long period of time as is input by weathering.

Weathering rate, removal rate and style of failure

In the preceding paragraphs, it was stated that rate of landsliding on a slope was governed by the rate at which weathering prepared material for removal. To a large extent this is true, with the proviso that transport systems are available at the base of the slope to remove slipped debris at a rate roughly equal to the rate at which it is produced by weathering. Thus, for a particular style of landslipping to be an equilibrium process, there must be a balance in the weathered zone such that rate of accumulation through time is balanced by rate of removal by slipping and transport away from the slope system.

Let us consider a situation where material is removed only slowly from the slope after a slip has taken place. In this case, the debris will remain to inhibit further weathering by mantling part of the slope and will preclude slipping by reducing the slope angle. The system will therefore be forced to adjust to the lower removal rate by a reduction in the rate of slipping or a change in process. This may be achieved by increasing the interval between slips, and one way in which this may be brought about is by lowering the slope angle such that the regolith will need more weathering to render it unstable.

Now let us look at a slope where the rate of removal has increased, so that it now exceeds the rate at which regolith is produced by weathering. If this situation persists, all the weathered mantle will be removed from the slope to expose fresh bedrock to attack by removal processes. As an example one can think of a coastal cliff in overconsolidated clay where cliff retreat is rapid. In such a case, weathering does not have time to reduce the strength before failure and so the intact strength parameters will control stability. The style of failure therefore changes in order to adjust to the increased basal removal, and in the case of overconsolidated clay the change will be from shallow residual slips to intact-strength rotational slips.

Many more examples of the relationship between failure style and weathering/removal balance could be envisaged. Suffice it to say that in any given material the style of failure depends on the relative weathering/removal rate and that, if that ratio changes, style and rate of failure and slope angle will change in order to restore an equilibrium. In some cases, of course, the adjustment may lead to the cessation of landsliding altogether, when the slope angle falls below the critical angle for the regolith in its most weathered state.

Further Reading

BISHOP, A. W. (1973). 'The Stability of Tips and Spoil Heaps', *Quarterly Journal of Engineering Geology*, 6, 335–76.

CABRERA, J. G. and SMALLEY, I. J. (1973). 'Quick Clays as Products of Glacial Action: a New Approach to their Nature, Geology, Distribution and Geotechnical Properties', *Engineering Geology*, 7, 115–34.

CARSON, M. A. (1971a). Op. cit. page 56, chapter 4.

—— and PETLEY, D. J. (1970). 'The Existence of Threshold Slopes in the Denudation of the Landscape', *Transactions of the Institute of British Geographers*, 49, 71–95.

CHANDLER, R. J. (1970a). 'A Shallow Slab Slide in the Lias Clay near Uppingham, Rutland', *Géotechnique*, 20, 253–60.

—— (1972a). 'Periglacial Mudslides in Vestspitzbergen and their bearing on the Origin of Fossil Solifluction Shears in low-angled Clay Slopes', *Quarterly Journal of Engineering Geology*, 5, 223–41.

FARMER, I. W. (1968). *Engineering Properties of Rocks*, Spon, London.

HOEK, E. (1973). 'Methods for the Rapid Assessment of the Stability of three-Dimensional Rock Slopes', *Quarterly Journal of Engineering Geology*, 6, 243–56.

HUTCHINSON, J. N. and BHANDARI, R. K. (1971). 'Undrained Loading, a Fundamental Mechanism of mudflows and other Mass Movements', *Géotechnique*, 21, 353–8.

KENT, P. (1966). 'The Transport Mechanism in Catastrophic Rockfalls', *Journal of Geology*, 74, 79–83.

MCGOWN, A., SALVIDAR-SALI, A. and RADWAN, A. M. (1974). 'Fissure Patterns and Slope Failures in Till at Hurlford, Ayrshire', *Quarterly Journal of Engineering Geology*, 7, 1–26.

SKEMPTON, A. W. (1964). 'The Long-Term Stability of Clay Slopes', *Géotechnique*, 14, 77–104.

SMITH, G. N. (1974). *Soil Mechanics for Civil and Mining Engineers* (3rd edition), Crosby, Lockwood, Staples, London.

TERZAGHI, K. (1962). 'The Stability of Steep Rock Slopes on Hard Unweathered Rock, *Géotechnique*, 12, 251–70.

WENTWORTH, C. K. (1943). 'Soil Avalances in Oahu, Hawaii', *Bulletin of the Geological Society of America*, 54, 53–64.

5 Slow Mass Movement Processes

Like chapter 4, this chapter is concerned with sediment transfer processes in which grain-to-grain contacts are maintained throughout movement. But the mass movement processes under discussion here act at much slower rates. Slow mass movements may take place along a well-defined plane in the same way as a landslide, but it is much more usual for shear to be distributed through a layer of soil or regolith, rather than confined to one depth. Slow mass movements are usually so slow that the actual velocity of the processes is a controlling factor in the input/output balance of sediment from a slope or a soil cover. In landsliding, movement of the slide is so rapid that it is not an effective control of the rate of sediment removal. It is other rates, such as the weathering rate or downcutting rate, which control process in the case of landsliding. Slow mass movements generally cause no immediate visible evidence of their operation, such as the scars and cracks which are associated with slides. Indirect evidence may be available, whose reliability is discussed below (section 5.2).

Sharpe (1938) introduced the term 'soil creep' to describe the slow movement of a soil cover, imperceptible except to long-term measurements. Creep has since been applied as a term to most styles of slow mass movement in sediment. This is to some extent unfortunate because there is a wide range of materials and of processes involved in slow mass movement, and it is misleading to think of them under one heading which implies a specific type of motion. Some of the more common materials exhibiting slow mass movement are single rocks on bedrock slopes, screes and soils and weathered regoliths. But in addition, unweathered hard and soft rocks and glacier ice exhibit slow movements. Processes may be due to gravitational, climatic or biotic forces and may depend on a variety of sediment properties for their operation.

In this chapter emphasis is placed on creep processes which are effective in sediment transport. Accordingly little will be said about intact rock creep which is generally so slow that it is unimportant to the direct movement of sediment, although it may play an indirect role by reducing rock strength and preparing it for failure. This must be regarded as a generalization and there are many cases, especially in clays, where rock creep is sufficient to produce appreciable movement. We now consider two major creep processes.

5.1. Creep of rock particles on bedrock and scree slopes

The ability of a body to creep down an inclined surface due to temperature fluctuations was noticed many years ago by Moseley (1869) who recorded movements of up to 45 cm in two years in the lead sheet roofing of Bristol Cathedral! Basically, the mechanism proposed was that greater downslope expansion and smaller upslope contraction in a body caused a net shift in position throughout a daily temperature cycle (Fig. 5.1). Little has been added to the theory of *insolation creep* since Moseley's work, which included an empirical relationship for movement during one temperature cycle of range *T*:

$$d = \frac{alT \cdot \tan \alpha}{\mu}$$

where *d* is the net movement of a body whose coefficient of linear expansion is *l*, and coefficient of sliding friction μ (see Fig. 5.1 for other terms). Basically, displacement is proportional to body length and inversely proportional to friction at the contact.

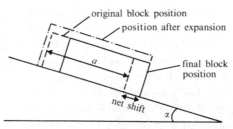

FIG. 5.1. The creep of a body down an inclined plane due to temperature fluctuations.

The insolation creep mechanism was first considered possible for the creep of rock particles on bedrock and scree by Davison (1888) who carried out controlled experiments on the process. Using one cut stone block resting on another, Davison noted movements of up to 13·17 mm/yr on a 17° slope. Days with frequent temperature fluctuations gave the fastest rates of movement, demonstrating the importance of insolation in the mechanism. Tamburi (1974) has studied rock creep under laboratory conditions and found that the rate of movement was proportional to the sine of the slope angle, grainsize and temperature, and inversely proportional to the coefficient of friction. These findings are very similar indeed to those deduced by Moseley for lead sheet creep. Tamburi (1974) also studied natural rates of insolation creep in the Colorado Front Range district of the United States. He monitored the movement of ten large boulders on bedrock surfaces and found rates of shift of between 0·16 and 2·1 mm over one year. Most of the movement could be explained by sine of slope and slope aspect, of which the most important factor was slope sine. The relationship with aspect is due to different temperature ranges on slopes facing in different directions. It should be noted that the blocks used were very large, weighing from 600 kg to 6500 kg, and were probably unaffected by any process other than insolation creep. Tamburi found that particles could either move in the simple way envisaged by Moseley or by the 'stick-slip' mechanism. Stick-slip is a sudden jerky movement in response to a build-up of stress, followed by sudden cessation of motion. There did not seem to be any simple explanation why one type of movement occurred and not the other during any experiment, though there was some tendency for stick-slip to be favoured on steeper slopes.

Schumm (1967) found that annual movement of rock fragments on shale slopes (angle α) in Western Colorado could be approximated by $d \approx 100 \sin \alpha$. The mechanism here was freeze–thaw action similar to that envisaged by Davison (1889) (section 5.2). Particles were raised normal to the ground surface on freezing but fell back vertically during thaw, aided by reconsolidation of the frost-heaved surface by rainbeat. Williams (1974) has subsequently found similar process rates on slopes subjected to seasonal slopewash. Undermining of the rock particles by flowing water was the main cause of movement.

Creep of the surface layer of a scree slope has long been suggested as a significant particle movement process on scree slopes. The main difference between creep on bedrock and creep in the surface layer of a scree has been highlighted in a theoretical discussion by Scheidegger (1961). In insolation creep, the whole surface of a scree is free to move, whereas only loose particles can do so on a bedrock surface. If the whole surface layer expands laterally during heating, the strength of the sediment is exceeded by stresses induced causing shearing and overriding of particles in the surface layer. The effect would be similar to buckling of railway lines due to heating if no expansion gaps are left between the rails. There are no measurements of insolation creep velocities on scree slopes, but it does seem likely that the velocity will be lower than rates of single particle creep on bedrock surfaces, because scree particles interlock with each other and constrain motion. Using marked stones, Caine (1963) measured rates of scree particle movement of 5–40 cm/yr on fine-grained scree in the Lake District, in which appreciable interstitial water could be held. Movement, almost entirely confined to winter, was due to frost heave. Rapp (1960) has also recorded movements of 0–22 cm/yr in Northern Sweden and found little relationship with slope gradients. High rates of surficial scree particle movement have been measured by Gardner (1969) who found annual displacements of up to 41 metres, with over a third of the stones showing no movement at all. It is doubtful if much of this was due to creep. Impact by falling particles and other disturbances were almost certainly the major cause, and this supposition is supported by the fact that there is increasing movement towards the top of scree slopes where impacts would be most numerous. It is probable that creep on scree slopes is a very slow process indeed, unless enough interstitial fine material is present to retain water and therefore to propagate frost heaving.

There are very few field measurements by which we can judge the efficacy of rock particle creep in sediment transport, but enough exist to show us that rates are low. If we return to our 30° slope of 100 metres' length, a boulder being shifted by insolation creep would take a minimum of 47 600 years to reach the bottom, on measured rates. Similar calculations of velocity for the freeze–thaw mechanism of a rock creep reported by Schumm (1967) yield a minimum travel time of about 1450 years, and for freeze–thaw in fine-grained screes a minimum travel time of 250 years is about right. The important question is whether the process acts sufficiently quickly to cause significant over-all movement of material, or whether other processes generally act faster and obscure the effects of rock creep. The blocks monitored by Tamburi (1974) were so large that it is unlikely that any process other than insolation creep could effectively move them. However, the rates are so slow that there is probably time for *in situ* weathering to reduce boulders to small particles, which could then be moved by other processes. Freeze–thaw action produces faster rates of rock creep than insolation creep, provided conditions favour it, and particles may move 100 metres in 250–1500 years. The rates reported were for relatively small rock fragments. Again, weathering may break down the particles as they move to a more easily transportable size, but smaller rock fragments may already be susceptible to movement by flowing water on the slope if it occurs. The frequency of critical flows then becomes a controlling factor in whether rock creep is effective or not. Once picked up by a flow, particles could be transported from a slope in geomorphically instantaneous time, and so the importance of rock creep decreases with particle size and with increasing recurrence of effective slopewash. The fact that a boulder may creep a few metres before disintegrating to easily transportable particles or before it is picked up by slopewash is irrelevant to its effective rate of transfer from the slope. Other processes, such as slopewash, move sediment at relatively high velocities. So-called boulder-controlled slopes of semi-arid areas indicate the importance of breakdown during transport as a process control. Boulder-controlled slopes are common in semi-arid areas and consist of a thin veneer of debris in transport from a cliff face over a bedrock slope. Considerable evidence exists to show that grainsize declines downslope on these slopes, showing that particles disintegrate as they progress downslope. In the lower part of the slopes, effective debris transport is almost entirely by flowing water and not by creep, due to the small particle size.

5.2. Creep in soils and regoliths

Slow, almost imperceptible creep in soils and regolith over bedrock slopes has long been recognized as possibly important in the mass transport of sediment. Slow mass movements, unlike their rapid counterparts, act continuously and over wide areas, which tends to compensate to some extent for their slow rate of operation. It is only recently that techniques have been developed to measure the rather slow rates of creep in soils, and hitherto movements were detectable only by their indirect effects on external bodies. Sharpe (1938) lists and illustrates a wide range of observable displacements of objects which he assumes to be attributable to soil creep. The most important of Sharpe's creep indicators are:

(1) Curved tree trunks—trees attempt to grow vertically but are continuously displaced from the vertical by soil creep. Since subsequent growth tends to be vertical, a convex downslope curve to the trunk should develop through time. There is also a tendency for the root system to trail upslope from the tree because it becomes anchored in slower-moving soil close to the bedrock.

(2) Displaced bedrock—a number of displacements in bedrock are ascribed to soil creep. For example, downslope drag due to soil creep tends to bend underlying strata, and fragments from more resistant beds or veins are swept downslope to form a tail of particles in the creeping soil.

(3) Displaced man-made structures—fence posts, walls, roads and many other structures may be shifted out of position by creep.

The validity of qualitative soil-creep evidence has been questioned in recent years. In particular, evidence from curved trees has met with criticism on a number of grounds, which have been reviewed and investigated by Phipps (1974). First, a tree subject to steady tilting by creep should form a continuous curve from ground level to canopy but it is all too frequently found that curvature is confined to a single bend close to ground level. Secondly, tree curvature is often confusing, with trees on the same slope bending different ways and other trees bending on no slope at all. Thirdly, investigations of tree-root systems have found no support for the suggestion that root systems trail upslope. Alternative mechanisms for tree curvature will not be discussed here but are treated at length by Phipps (1974), who found no evidence to suggest that curvature was due to soil creep. Displacement of man-made objects is undoubted proof of local soil movement, but this should be used guardedly as evidence for more widespread movement. The objects impose an extra load on a small area, which may be sufficient to generate local motion in what is a generally static soil. Bedrock curvature beneath a soil is probably a manifestation of soil creep, provided curvature increases steadily towards the bedrock/soil interface. Frequently, strata bend very sharply in a narrow zone near to the ground surface, which is probably indicative of toppling or collapse of the bedding planes in a downslope direction. It is possible to be sure of soil creep movements only by making direct measurements, and even then only within the limits imposed by accuracy and soil disturbance due to installing the measuring devices. Measurement requires the insertion of reference points into the soil surface or throughout the entire profile, whose initial positions are accurately known. Re-measurement of their position at a later date shows how much movement is taking place.

Creep movements in soils are caused by many mechanisms but are generally a result of one or more of three major types, viz. heave and settlement, downslope stress alone, and random particle displacements with a net downslope component. Although it is quite possible to distinguish between mechanisms in a theoretical sense, it is often very difficult to say how much movement is attributable to each in a real soil. However, the major creep mechanisms are:

(1) Creep in response to downslope gravitational stress—*Continuous creep*
(2) Freeze–thaw ⎫ expansion (heave) and contraction (settlement)
(2) Insolation ⎬ mechanisms due to climatic factors—*seasonal*
(4) Moisture changes⎭ *creep.*
(5) Collapse of animal burrows, root holes and other voids with a net downslope shift of particles—*random creep.*

Although continuous, seasonal and random creep processes may take place in a soil simultaneously, we shall consider each as a distinct mechanism and assess them separately. Faster rates of soil creep in periglacial environments, termed *solifluction*, are treated as a separate issue and the extent to which mechanisms compare with soil-creep processes is discussed.

Continuous creep

Continuous creep (Terzaghi, 1950) refers to slow, steady movement of a soil cover entirely in response to downhill gravitational stress. It is a feature of materials in which strain is a function not merely of stress but also of time. Many natural and artifical materials such as ice, rock, soil, and metals, undergo creep but we shall mainly confine our attention to soils. The general response of a material to stress is a complex interrelationship between applied stress, strain and time, in which knowledge of two quantities can be used to predict the state of the third. The form of the relationship depends on a soil's rheological characteristics. It may be recalled that it is usually adequate to think of soils as plastic materials with respect to landsliding. That is, strain consists of a very rapid displacement along a plane where shear stress exceeds shear strength. Time is irrelevant, in so far as strain is very rapid and only occurs when shear strength is exceeded. In some soils, particularly clays, slow strain or *creep* occurs when applied stress is lower than the yield stress at a rate which is stress-dependent. This is usually interpreted as an element of viscous behaviour in the soil's response to stress. The principle is illustrated in Fig. 5.2. Sandy soils are almost entirely plastic in deformation, failing above a well-defined yield stress. Clays also possess an identifiable yield strength at high stresses, though when this is exceeded, the rate of strain is to some extent stress-dependent. At lower stresses in clays, however, deformation still continues at slower rates, reflecting the viscous character of the material. Ice is not dissimilar in its stress response except that, unlike soils, it continues to deform down to zero stress.

The rheological response of a soil (behaviour under stress) is measured using a modified, triaxial, test procedure called a *creep test*. Instead of increasing the axial load throughout the test the stress is held constant and the

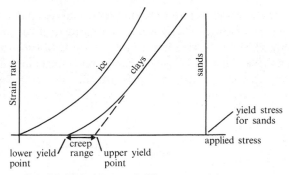

FIG. 5.2. The response of different materials to stress.

rate of deformation measured. The results of many creep tests on clays allow
certain generalizations to be made. Rate of strain tends to decline through
time and approaches a constant level, which at low levels of applied stress
may be zero. The precise behaviour depends on the level of applied stress
and also on water content. Clays can exhibit a range of behaviour from
almost purely elastic or plastic through to almost purely viscous, as these
two parameters vary. Generally if applied stress is less than residual strength
of the clay, creep rate declines through time to zero and is not significant in
long-term behaviour. At stresses lower than the yield stress but higher than
residual strength, deformation approaches a constant rate through time. In
this stress range, however, there is the probability of progressive failure to be
taken into consideration. That is, through time, strength falls to the residual
level due to loss of cohesion and fails rapidly. Thus the duration of continuous creep is limited by the length of time required for strength to fall to
residual levels. This itself is stress-dependent; the higher the applied stress
the shorter the duration of steady-rate creep. Above the yield point, of
course, failure is rapid and creep insignificant. These generalizations of behaviour are summarized in Fig. 5.3. From this discussion it is apparent that
continuous creep requires a delicate force-balance to operate over long peri-

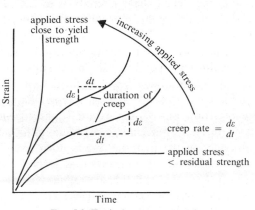

FIG. 5.3. Typical creep test results.

ods of time. Below residual strength, it stops quickly. Above, it eventually accelerates towards failure. Therefore, only at stresses just greater than residual strength will it continue for very long periods of time (Kirkby, 1967).

Velocities and velocity profiles have rarely been measured in soils suffering continuous creep, but two examples are illustrated in Fig. 5.4a. Both exhibit movement to appreciable depths (up to 10 metres) at quite high rates of movement. Maximum velocities are up to 130 mm/yr and transport rates of up to 2500 cm^3/cm/yr (cubic centimetres per centimetre width of slope per year) are obtained. These should be compared with rates of seasonal creep discussed below. Theoretical velocity profiles for a visco-plastic, solid, undergoing creep have been calculated by Yen (1969). He derived the following formula for velocity (u) at depth x in a soil of thickness h:

$$\frac{u_x}{u_{max}} = 1 - \frac{(x - x_{max})^2}{(h - x_{max})^2} \cdot$$

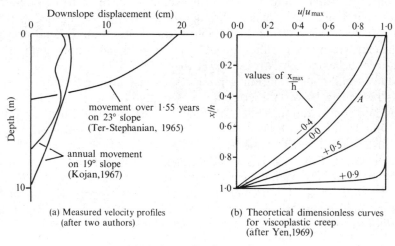

(a) Measured velocity profiles (after two authors)

(b) Theoretical dimensionless curves for viscoplastic creep (after Yen, 1969)

FIG. 5.4. Velocity profiles for continuous creep.

Some theoretical profiles are shown in Fig. 5.4b and, for a soil with no residual cohesion, maximum velocity is at the surface, and the profile corresponds to curve A in the diagram. The measured and theoretical profiles are broadly similar, having a convex downslope form with most creep concentrated in the lower part of the profile. As an example, in the case reported by Ter-Stephanian (1965). (Fig. 5.4a), over half of the movement is in the lowest quarter of the soil.

Seasonal creep and solifluction

Seasonal creep processes are not so closely related to specific soil-types as continuous creep, and can take place in any debris which undergoes cyclic volume changes. Mechanisms responsible are generally related to climatic processes which cause a heave parallel to the soil surface and a near-vertical settlement on contraction. A net downslope displacement of particles in the expanding and contracting layer results from this, as we shall see in more detail below. Freeze-thaw temperature oscillations, daily heating cycles and moisture-content changes in the soil have all been cited as mechanisms producing cyclic volume changes in soils. Of these the latter is the one most confined to specific soil types. It is most effective in fine-grained soils where unsaturated conditions generate large suction forces drawing the particles closer together, and in soils containing clay minerals which swell appreciably on absorbing water (e.g. montmorillonite). Nevertheless, repeated soil-moisture content change is probably a more widely distributed process in soils than either freeze–thaw oscillation or high diurnal temperature fluctuations, and in that sense may be globally more important. It is of major significance that seasonal creep mechanisms are not dependent on a very delicate force-balance for their operation, unlike continuous creep. There is potential for displacement on any inclined surface, no matter how low its angle, and in a very wide range of soil types. Consequently, seasonal creep approaches continuous-ubiquitous operation more closely than any other process.

Davison (1889) was among the first to postulate the heave and settlement mechanism due to freeze-thaw as a means of movement in regoliths. He proposed that expansion on freezing would be more or less normal to the ground surface, whereas settlement would follow an almost vertical path. The theoretical particle path (Fig. 5.5a) is a simple zig-zag, where net displacement is proportional to slope angle. For such a path to be followed, soil cohesion must be so great as to preclude movement under gravitational forces during expansion, and yet negligible in contraction so that settlement is vertical. Kirkby (1967) modified Davison's approach to take into consideration the forces due to overburden pressure and soil cohesion in the regolith. A force diagram at a point in the soil is shown in Fig. 5.5b. Cyclic forces, acting at an angle to the soil surface, are assumed to be exactly opposite in direction in expansion and contraction. A vertical force due to overburden is present, and the resultant direction of motion is indicated by R. Particle path is therefore inclined to the downslope side of normal during expansion, but to the upslope side of vertical in contraction. Davison's assumption concerning great cohesion in expansion and negligible cohesion during contraction is perhaps not all that unreasonable for freeze-thaw, because the presence of ice would increase soil strength in expansion and meltwater might decrease it in contraction. Davison's assumptions are not justifiably applicable to expansion and contraction cycles due to moisture-content variations, and so one must take soil-strength factors into consideration. Measured particle paths (Fig. 5.5c) on controlled blocks of soil approximate fairly closely to the general zig-zag pattern proposed by Davison and modified by Kirkby. The tendency for expansion and contraction to be displaced from normal and vertical directions respectively is well shown by measurements of displacement of experimental soil blocks. Movement near boundaries of a soil block tended to be different. On the upslope boundary, net movement was upslope, and on the downslope boundary, downslope movement was enhanced. Although this is undoubtedly an edge effect, due to the experimental procedure, internal boundaries exist in soils in the form of fissures and often at a spacing close enough to be significant. Fleming and Johnson (1975) found that displacement at any point in a real soil was strongly controlled by the proximity of fissures, and it is important to note that movement can show considerable spatial variation in direction and magnitude on the same slope, even to the extent that upslope movement is possible locally.

Davison (1889) expected that a single freeze-thaw cycle would cause a linear soil displacement, with the amount of displacement declining to zero at the limit of freezing. That is to say, displacement would be directly proportional to the distance from the base of the freezing layer. But, as Davison pointed out, individual freezing cycles vary in temperature range and depth

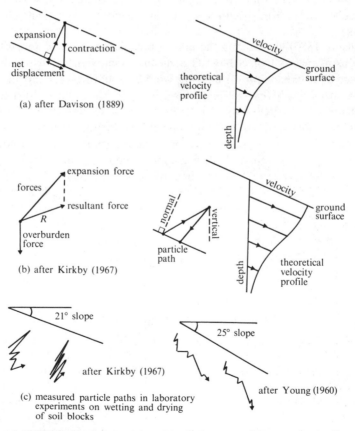

Fig. 5.5. Theoretical and measured particle displacements due to seasonal soil creep.

of penetration into the regolith, resulting in very high rates at the surface where *all* cycles have some effect, and dying away rapidly to zero at the depth where no freezing ever occurs. Consequently, a concave downslope velocity profile is to be expected (Fig. 5.5a), assuming average heave decreases exponentially with depth. If soil cohesion and the overburden force are taken into consideration, a different pattern of movement results. Since, at the ground surface, overburden force is zero, the resultant particle movement (Fig. 5.5b) after a heave cycle is also zero. It is apparent, then, that the maximum zone of movement lies at a depth intermediate between the surface and the maximum depth to which the heave mechanism extends. A concave-convex velocity profile results from this with maximum shear somewhat below the surface (Fig. 5.5b), again assuming an exponential decline in average heave with depth. It can be seen that the over-all difference in velocity profile according to these two models is small, and confined to the uppermost part of the soil.

Measured velocity profiles on creeping soils show the general trend towards rapidly diminishing movement with depth which is a characteristic of theoretical profiles mentioned above. A number of profiles from a wide range of environments is illustrated in Fig. 5.6a. All show a rapid decay in creep rate with depth, but in detail are complex and do not show a good fit with theory. One curve in Fig. 5.6a shows two maxima, and this finding was attributed to different properties in the upper and lower soil horizons. We should not be too surprised that field creep measurements do not accord perfectly with the theoretical heave mechanism. Quite apart from the fact that the heave theory embodies assumptions about the soil and process which may not be entirely correct, other mechanisms, including continuous

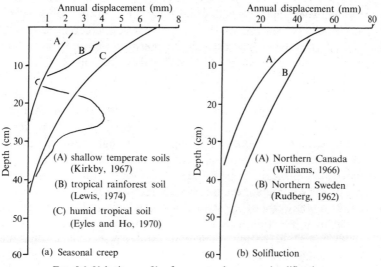

FIG. 5.6. Velocity profiles for seasonal creep and solifluction.

and random creep, may occur simultaneously with seasonal creep and will tend to complicate the picture. We should also not forget that measuring creep is difficult, inevitably involving some soil disturbance, and measured profiles may not themselves be entirely accurate representations of reality.

Depth of creep is clearly limited to the depth of penetration of the seasonal heave mechanisms. Freeze–thaw seldom penetrates more than a few centimetres in temperate latitudes (Caine 1963) and even in permafrost (permanently frozen ground) the layer of seasonal melting (active layer) seldom extends to more than a metre or two. Marked diurnal temperature fluctuation extends only a few centimetres into solid rock (Peel, 1974) and would extend a much shorter distance into soil whose thermal conductivity is lower. Finally, the soil moisture-change mechanism is restricted to the zone influenced by evapotranspiration and drainage. Usually this is of rather shallow depth, though exceptionally it may be many metres thick. Generally speaking then, seasonal creep should not be significant below depths of 20 cm to a metre.

Rates of sediment movement by creep are about 1–5 cm^3/cm/yr in temperate latitudes, rising to around 10 cm^3/cm/yr in both tropical and colder regions (Young, 1974). One point which does arise from the many studies quoted of soil creep is that the theoretical relationship between rate and slope angle has not been conclusively demonstrated in the field, even within quite small areas. One must conclude that soil factors and possibly other mechanisms obliterate any effect which topography might have upon the creep process.

Creep rates increase towards periglacial areas (permafrost zones), where the process is frequently referred to as *solifluction*. There is no sharp distinction between solifluction and seasonal creep rates, and considerable overlap occurs between the fastest seasonal-creep and slowest solifluction rates. However, certain differences other than rate between creep processes in periglacial areas and other climatic zones do merit attention. First, solifluction is not uniformly distributed spatially but tends to be concentrated in rather well-defined lobes separated by relatively immobile areas. Young (1974) has drawn attention to this problem, and advises caution in interpreting measured rates because observations are concentrated in the more mobile zones. Secondly, movement is also temporally restricted to the time of year when the active layer is thawed out. Consequently, although annual rates are from five to ten times higher than for seasonal creep (up to 80 cm^3/cm/yr) all the movement takes place in a short space of time at quite a high velocity. Mechanisms for solifluction include heave due to freeze–thaw and also flow along zones of thawed-out ice lenses in the soil. An alternative to the interpretation of solifluction as a creep mechanism was raised by Chandler (1972), who has observed low-angled mudslides in periglacial Spitzbergen. These have similar characteristics to solifluction lobes and are due to high pore-water pressures inducing shallow sliding. The high pore-pressures could have been due to thawing out in ice lenses, or possibly due to pressure exerted on the lower soil by overnight freezing at the surface. Whatever the cause, Chandler was able to explain movement within the lobes primarily in terms of shallow sliding. Fossil shear-planes on low-angled, and now quite stable, clay slopes in Central England have also been interpreted in the same way (Chandler, 1970). Velocity profiles (Fig. 5.6b) imply that by no means all solifluction lobes are interpretable as mudslides. They show the characteristic decline of movement with depth found in seasonal creep and do not demonstrate any particular tendency towards simple basal shear implied by a slide mechanism.

Random creep

The activity of many soil organisms, animal and vegetable, generates a net downslope displacement of soil. Burrowing animals produce cavities which tend to collapse through time and collapse will be preferentially in the downslope direction. The same applies to root channels. Worms and other small animals deposit much material at the surface in a loose state which will easily be moved in a downslope direction (section 3.3). It is difficult to assess the over-all importance of these processes, but they obviously contribute to creep rates measured in soils.

5.3. Conclusion

It is apparent from a discussion of slow mass-movement processes that they result from a very wide range of mechanisms operating in a very wide range of materials. To consider all these processes under the traditional heading of creep is misleading, and leads to erroneous implications about the way processes operate. Only one slow mass movement, continuous creep, can justifiably be regarded as a true creep process in the rheological sense. The majority are heave processes which are preferentially exploited in a downslope direction. Even the term 'slow mass movement' is open to criticism, however. For example, one could argue that continuous creep is not a mass movement but a flow, since it depends on a viscous component of soil behaviour. Also, on the basis of process and rate, solifluction may often be a slide process and should really be considered a rapid mass movement.

Further reading

CHANDLER, R. J. (1972a). Op. cit. page 100.

DAVISON, C. (1888). 'Note on the movement of scree material', *Quarterly Journal of the Geological Society*, **4**, 232–8.

DAVISON, C. (1889). 'On the creeping of the soil cap through the action of frost', *Geological Magazine*, **6**, 255.

KIRKBY, M. J. (1967). 'Measurement and Theory of Soil Creep', *Journal of Geology*, **75**, 359–78.

PHIPPS, R. L. (1974). 'The Soil Creep—Curved Tree Fallacy', *Journal of Research of the United States Geological Survey*, **2**, 371–8.

RUDBERG, S. (1962). 'A Report on some Field Observations Concerning Periglacial

Geomorphology and Mass Movements on Slopes in Sweden', *Biuletyn Peryglacjalny*, **11**, 311–23.

SCHUMM, S. A. (1967). 'Rates of Surficial Rock Creep on Hillslopes in Western Colorado', *Science*, **155**, 560–1.

TAMBURI, A. J. (1974). 'Creep of Single Rocks on Bedrock', *Bulletin of the Geological Society of America*, **85**, 351–6.

YEN, B. C. (1969). 'Stability of Slopes undergoing Creep Deformation', *Journal of the Soil Mechanics and Foundation Engineering Division, Proceedings of the American Society of Civil Engineers*, 1075–96.

YOUNG, A. (1974). 'The Rate of Slope Retreat', *Institute of British Geographers, Special Publication No. 7*, 65–78.

6 Material Transport in Fluids

6.1. Introduction

The major distinction between fluid transport and mass transport is that in a fluid medium sediment grains are separated. This results in a reduction of the importance of grain-to-grain contact and interaction to small proportions. Sediment transport in fluids also tends to be selective. There is a critical size of particle greater than which the flow is incapable of entraining. This is normally referred to as *competence* of the flow. The selectivity of flow processes is also reflected in a tendency for smaller particles to be transported further and more quickly. Hence, unlike mass transport, fluid processes sort sediment grains by size in the direction of flow.

It is as well to emphasize here the importance of solutional transport in water flow systems. Mass movement processes carry only solids but, in contrast, large quantities of material are carried in solution by rivers and streams. Furthermore, solutes move as an integral part of the water, at the same velocity. The mechanics of solute transport are therefore identical to those of water movement. Solute movement occurs at all velocities (there is no critical velocity) and total load is easily assessed as the product of concentration and discharge. Solute concentration is related to the interfaces over which the water has flowed and the length of time in contact with them as well as to the mixing of different water bodies. It tends to be roughly constant with depth in the flow, unlike suspended sediment concentration which is very variable.

Water on the earth's surface is involved in the hydrologic cycle, falling as rainfall on to the land surface and making its way back to the oceans by various routes, including soil and groundwater seepage and overland flow. The land-based part of the hydrologic cycle is extremely complex but its overriding feature is a tendency towards increasing concentration along the system. A diffuse, ubiquitous input (rainfall) is combined into a discrete output at the river mouth. The extent of this concentration may be appreciated from the fact that nearly half the area of England and Wales drains to just four outlets, the Humber, Thames, Severn and Wash. In areas of permanent rivers, the development of well-integrated channel networks defines areas of land called drainage basins which drain to common outlets. All water and derived sediment and solutes must pass out of the drainage basin via the single outlet, and so the drainage basin provides a very convenient and meaningful transport system (chapter 7).

One must not forget the existence of other transporting fluids. The atmosphere, for example, is a fluid and is a very important transporting medium over large parts of the earth's surface not protected by vegetation. Wind behaves similarly to flowing water, except that air is about a thousand times less dense than water and is therefore less able to support sediment grains. Winds are almost entirely related to global air movements, and are only locally modified around topographic features. Consequently, there is no tendency for winds to concentrate along the direction of flow, and develop an integrated transport network in the same way as rivers. Analogies have also been drawn between the way glacier ice moves and fluid behaviour. The analogy is by no means strictly correct, however, and this is discussed in section 6.6.

The object of this chapter is to discuss the general underlying principles of sediment movement in fluids, and to stress the similarities of fluid transport in all environments. To this end, an introduction to the theory of fluid-transport mechanics is given in section 6.2. Since space precludes a lengthy discussion of specific fluid-transport systems, characteristics peculiar to particular environments are suppressed in 6.2 and are briefly discussed in subsequent sections of the chapter.

6.2. Mechanics of fluid flow and sediment movement in fluids

The mechanics of fluid flow and sediment transfer in fluids are much less easy to appreciate than the elementary static problems introduced in section 1.4, and extended in chapter 4. Fluids are in constant, irregular motion and consequently it is difficult to be precise about the forces exerted on fluid boundaries and on objects by the constantly changing flow. However, transport in fluids can be resolved into three main issues:

(1) Movement of the fluid—This is important because the velocity profile with depth in the fluid determines the force exerted at the bed.
(2) Boundary or bed stress—The stress exerted at the bed of a flow is the force per unit area available to move sediment particles on the bed. It should be noted that the stress exerted at the bed is exactly equal to the frictional resistance to the motion of the fluid.
(3) Transfer of energy and forces within the flow—Once sediment particles have been entrained in a flow, the force which resists their tendency to settle back onto the bed is made up of the forward motion of the fluid and of transfers of energy within the flow. The latter consist of turbulent transfers of the fluid and of collisions between grains and with the bed.

These three issues are discussed below.

Mechanics of fluid motion

Water and air can be considered perfectly viscous materials (chapter 2); that is, they move at any level of applied stress. In the same way as a sliding, solid body, a layer of fluid flowing over a surface is subject to frictional resistance. In a fluid, however, resistance to motion along internal boundaries is much less than at the bed because intermolecular forces are smaller. Hence each successive layer of fluid away from the bed is able to slip past the one below, distributing shear throughout the full depth of flow. Upper layers move faster than lower layers, giving a *profile* of velocity in the flow. In the type of flow described as *laminar flow*, layers of fluid glide past each other in parallel fashion and no mixing occurs between them (Fig. 6.1a). When flow velocity exceeds a critical value, dependent on other factors such as flow depth and bed roughness, laminar flow becomes unstable, and intermixing of layers begins. The simple pattern of laminar flow breaks down into *turbulent flow* and the parallel streamlines are destroyed by eddying. Shear becomes proportionately greater closer to the bed, and less away from it, owing to the mixing of slower bodies of water up the profile and faster ones down, tending to distribute velocity more evenly (Fig. 6.1b). Turbulent flow, however,

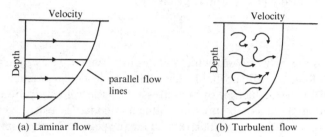

FIG. 6.1. Laminar and turbulent flow.

fluctuates considerably at a point and between points in an unpredictable fashion, rendering a rigidly determined velocity profile based on mechanical principles impossible to achieve.

Some impression of whether flow is laminar or turbulent can be gained from an empirical parameter called the Reynolds Number (N_r). For flow through channels N_r is given by:

$$N_r = \frac{Ru}{v}$$

where u is the fluid velocity and v its kinematic viscosity. R is the hydraulic radius of the channel, defined as:

$$R = \frac{A}{2D + w}$$

where A, D and w are area, depth and width of the channel respectively. It is apparent that R approaches depth of flow as the channel becomes very wide. Flow is laminar when R is less than 500, turbulent when R is greater than 2500, and between these values either type may occur.

Initiation of sediment transport in fluids

Grains on the bed of a fluid are picked up by the flow when the shear stress applied to them is greater than their resistance to motion. The first problem is to determine the shear stress at the bed due to the flow (bed stress), since this is an important driving force in grain movement. It may be derived theoretically from the velocity profile which Newton deduced for laminar flow as:

$$\frac{du}{dz} = \frac{\tau}{v\rho_f} \quad \text{or} \quad \tau = v\rho_f \cdot \frac{du}{dz} \tag{6.1}$$

where τ is shear stress at depth z, du/dz is the velocity curve gradient at depth z, and v and ρ_f are the *kinematic viscosity* and density of the fluid respectively. In turbulent flow this must be modified to account for the fact that not only is momentum transferred from layer to layer in the fluid but bodies of water are also transferred. By analogy with formula (6.1) above, shear stress in turbulent flow is given by:

$$\tau = \varepsilon_m \rho_f \cdot \frac{du}{dz} \tag{6.2}$$

where ε_m is the *eddy viscosity* which relates to turbulence as well as kinematic viscosity. Readers will probably have already foreseen practical problems arising from the deduction of bed stress from the velocity profile. Not the least among these is obtaining a value for eddy viscosity, which is dependent on flow factors rather than fluid properties. In the final analysis, however, application of the velocity profile to the derivation of bed stress is limited by measurement difficulties. First, for equation (6.2) to be applied requires a good knowledge of the velocity curve gradient at the bed, which is difficult if not impossible to obtain. Secondly, natural flow tends to fluctuate so widely at a point that it is not possible to measure a representative velocity profile with any certainty.

An alternative approach to the initiation of sediment transport may be found in the realms of statics. The total tractive force (T) exerted on the bed of a flow is given by:

$$T = \bar{\tau}_0(2D + w)L$$

where $\bar{\tau}_0$ is the average boundary shear stress (stress at bed and banks) and D, w and L are the depth, width and length respectively of a section of channel. (Note that this formula is for flow in channels. For unconfined flow, the depth term disappears.)

In a *steady*, non-accelerating flow the total force exerted at the bed is equal to the downslope component of the weight of water (W) in the reach:

$$W = \gamma_w DLw \cdot \sin \alpha = \bar{\tau}_0(2D + w)L$$

where α is the channel slope angle and γ_w is the unit weight of water. This simplifies to:

$$\bar{\tau}_0 = \gamma_w Rs$$

where s is the channel gradient (approximated by $\sin \alpha$ for low gradients) and R is the hydraulic radius. Thus, we have a very simply derived value for average bed stress, without reference to the velocity profile, and from this it is possible to analyse the critical condition necessary to cause sediment motion. It is necessary, however, to make assumptions about grain shape and packing.

For spherical grains of diameter d and submerged unit weight γ_{sub}, submerged weight is given by:

$$\gamma_{sub} \cdot d^3(\pi/6).$$

Movement of these grains will begin when a critical shear stress ($\bar{\tau}_{crit}$) is applied to them, given by:

$$\bar{\tau}_{crit} = n\gamma_{sub}d^3\left(\frac{\pi}{6}\right) . \tan \varphi$$

where n is the number of particles per unit bed area and φ is the angle of friction between particles and bed, either sliding or rolling friction according to the mode of movement. In terms of factor of safety (chapter 4), the safety factor (F) against grain movement is given by:

$$F = \frac{\text{resisting force}}{\text{driving force}} = \frac{n\gamma_{sub}d^3\left(\dfrac{\pi}{6}\right) . \tan \varphi}{\gamma_w Rs}.$$

The practical drawbacks with this method of analysis are numerous, and relate to the assumptions made about the sediment and the flow. Natural sediments are variable in grainsize and packing, and their constituent grains are irregularly shaped. $\bar{\tau}_0$ is the *average boundary shear stress* only. Stress varies considerably across a channel and will be equal to the average value in very few places. Pulsations in flow due to turbulence complicate the issue further by concentrating stress at some points but not at others. Measurements have shown that bed stress varies up to twice average during pulsations, which accounts for the observation that sediment movement often begins at lower than predicted bed stresses. Another point which is often overlooked is that fluid erosion usually begins during unsteady flow, for example when discharge and velocity are increasing due to a flood pulse in a river channel. All these complicating factors are difficult to accommodate theoretically and render the application of simple static models to fluid transport practically impossible.

A further complication relating to the flow is the presence of the laminar sub-layer adjacent to the flow boundary. If particles do not project beyond the laminar sublayer, the boundary is considered hydrodynamically smooth, and rough if the turbulent layer is penetrated. Flow and sediment movement in the laminar layer are poorly understood, but it is clear that greater shear stress is exerted on grains which penetrate the turbulent layer than on those which do not. Greater velocities are required to ensure that successively smaller particles penetrate the turbulent layer and hence be exposed to the higher shear stresses that exist there. This is well shown by Hjülstrom's (1935) now classic diagram based on experiments relating grainsize to critical erosion velocity in water (Fig. 6.2). Material in the fine to medium sand range is most easily moved, whereas smaller and larger grains require higher velocities to initiate transport. Cohesive forces between smaller clay particles also increase their resistance to erosion.

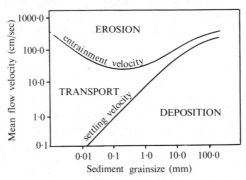

FIG. 6.2. Hjülstrom's critical erosion velocity diagram. (After Hjülstrom, 1935.)

So far we have neglected the importance of another force acting at a flow-bed and have considered only the bed stress exerted by the flow. There is a tendency for grains to be subjected to a vertical lifting force due in part to upward transfers of turbulent eddies from the stream bed, but also to the velocity gradient in the flow. A velocity gradient exists from top to bottom of grains at the bed, setting up a pressure gradient in the opposite direction and introducing a tendency for the grain to move vertically along the pressure gradient. This force, acting vertically from the bed, is called the *lift force*. There has been much discussion about the importance of lift force in sediment movement but measurements have shown that it may be up to two or three times as large at the bed as bed stress in some cases. It does however diminish rapidly away from the bed as the gradient of velocity declines (Fig. 6.1).

Thus, once a particle has been lifted into the flow, the lift force rapidly diminishes to insignificant levels, and it falls back to the bed. Observations made in flumes on grain transport tend to support the sequence of events suggested above. Particles have been seen to move vertically at first into the fluid, presumably by the lift mechanism, and then to move rapidly downstream as they become entrained in the flow. As they move forward however, they fall steadily back towards the bed because lift force is small within the fluid itself. This results in particles moving in a series of asymmetric hops along the bed (Fig. 6.3), a process known as *saltation*. Turbulent eddying is also very important in lifting particles from the bed, since the effect of turbulence is to transfer faster-moving water to the bed from higher

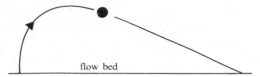

FIG. 6.3. Observed particle path in fluid transport.

parts of the flow. This is often seen as small whirls or eddies in the flow which suddenly whisk material from the bed.

Maintenance of sediment transport in fluids and rates of transport

Once sediment movement has been initiated, the forces acting on the grains in order to maintain motion may change, due either to variations in the lift force or to fluctuations in turbulence. In addition, the mode of particle movement is important in resistance to motion, whether it be in close contact with the bed in rolling or sliding, intermittent contact in bouncing, or totally suspended above the bed. Clearly, there are no sharp distinctions between these modes of transport, and bed movement tends to merge through saltation, to suspension of grains within the fluid. Nevertheless, they form convenient distinctions within which to discuss sediment movement. Much of the theoretical and experimental work on sediment movement and rates of transport has been carried out by hydraulics engineers concerned with problems such as the design of stable beds and banks for canals and irrigation ditches, and this is especially true of work done on *bedload*, which is considered first below.

Bedload comprises material which slides or rolls along the bed in response to the applied shear stress, due to fluid flow, and the downslope component of the material's own weight. The latter is usually small in most flows, but may become significant on steeper slopes. Resistance to motion is derived from friction and cohesive bonding between grains (Fig. 6.4). The traditional model for bedload movement follows the approach of DuBoys, proposed nearly 100 years ago. DuBoys assumed that sediment moved in response to bed stress as a series of parallel layers sliding over each other at progressively slower velocities with depth into the bed. At some depth velocity of sliding became zero, defining the limit of sediment layer in transport. It has now been shown that bedload does not move in this way, but DuBoys's subsequent argument does not depend upon this mechanical

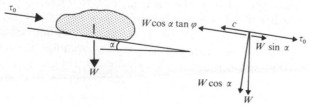

FIG. 6.4. Forces acting on particles in bedload transport.

model of sediment motion. Basically, DuBoys's approach to sediment transport was to equate shear stress at the bed with sediment transport rate. His formula shows that rate of transport is proportional to excess bed stress above a critical bed stress whose value depends upon the grainsize of the material in transport. The formula is of the form:

$$q_s = \chi\tau_0[\tau_0 - \tau_{crit}]$$

where q_s is sediment discharge in weight/unit width/unit time, τ_0 is bed stress, τ_{crit} is the bed stress required to cause movement of particles and χ is a constant dependent on flow parameters such as velocity and turbulence. Provided the flow is supplied with the maximum amount of sediment it is capable of transporting, q_s is the flow *capacity* and τ_{crit} a measure of its *competence* to transport the bed material. The problem with the DuBoys approach is in evaluating the constant, and many empirical studies have been carried out to determine it. Consequently, a large number of semi-empirical, DuBoys-type, bedload formulae exist. A variation of the approach is to relate transport to excess velocity above a critical level, rather than to bed-stress.

The DuBoys approach to bedload transport has provided some very successful empirical transport equations, but it suffers from serious theoretical and practical disadvantages. The formulae depend on determination of a critical bed stress (or velocity) for the initiation of sediment movement. It has already been shown that this is practically very difficult, due to the highly variable character of natural flow and bed sediments, and that theoretical critical bed stresses do not agree very well with experimental results. This objection has been overcome by an entirely different approach put forward by Einstein (1942, 1950). Observations in flumes have shown that particles do not move constantly along the bed, but that there is a continuous interchange of particles between mobile load and immobile bed. Particles move in rapid steps or jumps, with relatively long periods when they remain quite stationary. Furthermore, the steps appear to be independent of flow conditions and transport rate, and changes in rate can be effected by changing the period of inactivity between steps. Einstein's approach is an attempt to model this observed mode of transport, and therefore to evaluate the balance between particles leaving and returning to a unit area of bed in unit time. It is significant to note that no distinction is made between bedload and saltated (bouncing) load in this approach, and so the arbitrary distinction between the two is avoided. Of greater significance, however, is the fact that Einstein's approach does not in any way depend on consideration of a critical or threshold bed stress for erosion to begin. Entrainment of grains is more realistically assumed to take place during bed-stress fluctuations, perhaps due to turbulence and not due to the imposition of a steady, average stress. The approach utilizes the concept that there is a statistical probability of erosion taking place at any point through time, and represents a break with traditional mechanical models. In an equilibrium transport situation, the number of particles leaving a unit area of bed in a given time is equal to the number returning, and so neither net erosion nor deposition occurs. The transport rate therefore remains constant through time and can be analysed in terms of the number of particles passing through a cross-section of channel per unit time. This is equal to the number of particles dislodged from a section of channel one bounce-length upstream of the arbitrary cross-section, since those derived from further upstream will fall back to the bed before reaching the cross-section. The number of particles (N) leaving the section of bed one particle bounce-length upstream of the cross-section in unit time is given by:

$$N = \frac{Ap}{a}$$

where A is the bed area from which the particles are derived, p is the probability of any particle being removed per unit time and a is the bed area occupied by each particle. Accordingly, the transport rate (q_s) is:

$$q_s = Nv$$

where v is the volume of each particle. Both a and v are closely related to particle shape and size, and the theory is normally developed for spherical particles of a single grainsize or mixed grainsizes. Naturally, the drawback with Einstein's transport formula lies in the determination of p. Not unreasonably, this is taken as a function of the ratio between submerged particle weight and the instantaneous lift force at a point, which is effectively a function of the factor of safety against erosion.

A simple analysis of bedload movement by saltation was developed by Bagnold (1953), specifically to movement in air. It is based on the fact that a saltating grain loses momentum to the fluid because its 'take-off' velocity is lower than its 'landing' velocity. The rate of momentum loss (M) of particles whose initial and final horizontal velocities are u_1 and u_2 respectively is:

$$M = (u_2 - u_1)mN/Lt. \tag{6.3}$$

where m is the average particle mass, N is the number of saltating grains, L is the average bounce length and t is the time period. Transport rate (q_s) is the number of particles passing a point in unit time, given by:

$$q_s = Nm/t . \tag{6.4}$$

and combining equations (6.3) and (6.4) yields

$$M = q_s\frac{(u_2 - u_1)}{L} \tag{6.5}$$

Equation (6.5) is true, provided all grain movement is by saltation. At equilibrium, the momentum lost by the saltating grains must equal that imparted to them by the fluid. This is effectively the boundary shear-stress (τ_0) in air because most particles move as saltating grains. Also u_1 is effectively zero in the horizontal because initial movement is vertical under the influence of the lift force. Therefore, the transport rate is given by:

$$q_s = \frac{L}{u_2} \cdot \tau_0 \tag{6.6}$$

Of course, it is practically impossible to establish u_2 or τ_0 with any certainty, and so formula (6.6) above is of limited applicability. However, semi-empirical formulae have been derived for calculating transport rates due to saltation, which follow this approach. Space precludes further discussion here, but it may be followed up in sources quoted in 'Further Reading'.

The *suspended load* is supported entirely by the fluid and makes no contact with the bed during transport. Forces acting upon suspended particles, besides the general forward motion of the fluid, are twofold. First, there are upward and downward turbulent eddies. Over a time these should balance one another (otherwise the fluid density would be changing) giving a tendency towards uniform mixing of suspended and dissolved material through the flow. Secondly, the submerged particle weight is constantly acting downwards, which provides an over-all tendency for movement back towards the bed. The velocity of settling (V) is proportional to the square of grain radius (r) and for uniform spherical particles it is given by Stokes's Law:

$$V = {}^2\!/_9 \cdot (\rho_p - \rho_f)gr^2/v$$

where ρ_p and ρ_f are the fluid and particle densities respectively, v is fluid viscosity and g is the acceleration due to gravity. Owing to the tendency for solid particles to settle out of a flow at a velocity related to grainsize, the concentration and grainsize of suspended sediment increase towards the bed. With equilibrium conditions, the settling of particles on the bed must be balanced by upward turbulent transfer from the bed, such that an equilibrium profile of sediment concentration is established through the depth of flow. It can be shown theoretically that at equilibrium the concentration declines exponentially away from the bed. Transport rate in suspension is given simply by the product of average concentration and flow discharge.

It should be borne in mind that the rates of transport given by bedload, saltated load and suspended load formulae are maximum or *capacity rates*. For capacity transport, supply of sediment to the flow must be unrestricted. In fact, this is rarely the case and most natural flow is transporting much less material than it is theoretically capable of moving, simply because of lack of availability. Availability and derivation of sediment depend on many factors; these are described in the following sections which consider some aspects of transport in water, wind and ice.

6.3. Intermittent transport by water on slopes—wash transport

Rain falling upon the land surface is split into two parts, that which flows across the surface and that infiltrating and flowing through the soil. The surface-flow fraction forms an intermittent water-transport system whose important property is that there is no concentration of flow into specific flow-lines or channels through a long period of time. That is to say, flowing water moves as a continuous smooth sheet across the whole slope, or it gathers into small channels or *rills* whose positions shift from time to time, and therefore affects the whole slope over long periods. Its erosive effect is often called wash transport. This distinguishes hillslope-flow systems from other transitory systems, such as many semi-arid networks where stored water in the soil and groundwater are insufficient to maintain permanent flow. Once run-off begins on a slope it is fairly rapid in velocity to the extent that pick-up of solutes is small in comparison to subsurface flow (chapter 3).

Run-off may be generated in two circumstances, either when rainfall intensity (in mm/hour) exceeds the rate at which water can percolate vertically into the soil (infiltration capacity, also in mm/hour), or when it exceeds the rate at which water drains laterally through the soil. The former may be termed *infiltration excess overland flow* and the latter *saturation overland flow*. Infiltration excess overland flow is a more or less immediate response to heavy rain, and begins soon after rainfall begins, or after a time-lag if there are surface depressions on the slope which need to be filled with water before general flow can commence. Saturation overland flow, on the other hand, depends on the moisture content of the soil prior to rainfall, and there may be a considerable time-lag during which the soil fills up if it was initially quite dry. As a rough approximation, vertical infiltration rate I_v is a function of hydraulic conductivity (K), whereas the lateral or downslope rate I_l is a function of the product of K and hydraulic gradient, which is close to slope gradient ($\tan \alpha$) in the simple case of subsurface flow parallel to the ground surface. Also, there is an increasing tendency for saturation to occur downslope because of water draining from the upper part of the slope. Thus, the lower the slope angle the more likely that saturation overland flow will be generated because drainage is poorer. Actually the picture is complicated somewhat by the fact that I_v and I_l are also functions of time in that they tend to decline as rain continues. There are many reasons for this, for example, the compacting and sealing action of raindrop impact, and the closing of cracks and fissures as clay minerals absorb water and begin to swell.

Run-off causes an over-all lowering of the slope surface by various erosive processes known collectively as sheet erosion. First, the direct impact of raindrops (rainsplash) dislodges soil particles by flinging them into the air as parts of the drop rebound, and by pushing and undermining larger particles. Secondly, flowing water in a uniform sheet can entrain particles according to the principles outlined in the previous section. In this it is aided by increased turbulence due to raindrop impact with the flowing layer. Thirdly, water may concentrate into more or less well-defined flow lines, with a concomitant concentration of the erosive process. All these take place together—indeed cannot take place separately for any great length of time—and should be regarded as a composite, interactive process. This is exemplified by observations made by many workers, and it will serve as an example to consider some made by Moseley (1974) during experiments on rill erosion. Moseley's experiment was on an initially uniform, sloping, sand surface which was sprinkled with simulated rain. At first, all the rain infiltrated into the soil and no surface flow was generated, although some downslope soil movement took place due to rainsplash. After a time, puddles appeared as the soil filled up, and these coalesced to form run-off over the whole slope. The run-off soon collected into indistinct flow lines, with small, intervening areas where sheet flow continued to occur; these formed loci for the development of more distinct rills, fed by smaller rills draining from the interfluves. Simultaneously, over the whole surface, water was causing erosion in many different ways, emphasizing the complex composite nature of sheet erosion. These different aspects of a complex process are considered below.

Rainsplash

Raindrop impact moves material by a combination of rebound, pushing and undermining, as already mentioned above. On level ground, it achieves little more than a random rearrangement of particles, besides compacting the surface and making it more susceptible to run-off. On a slope, however, there is net transport of material due to a longer downslope 'flight path' for rebounded droplets and to the downslope component of the impact force (Fig. 6.5). The force exerted by raindrops is related to their momentum, the

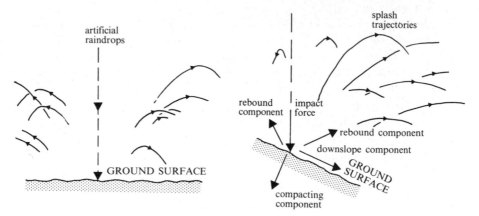

FIG. 6.5. Rainsplash trajectories. (Drawn from photographs in Ellison, 1948.)

product of their weight and impact velocity. Larger raindrops fall faster and so momentum may be expected to increase very rapidly with drop size and also with rainfall intensity, since larger drops occur in more intense rains (Table 6.1, Fig. 6.6).

The size of particles dislodged and their distance of transport by raindrops is thus closely related to momentum. In experiments carried out by Ellison (1945b) using raindrops of 3·5 to 5·1 mm diameter, 69–72 per cent of the material contained in rebounding droplets (that is physically thrown into the air) was less than 0·1 mm in size and 0·5 to 2·5 per cent was greater than 2 mm; soil contained in the run-off water was 87 to 92 per cent less than 1 mm and 0·01 to 0·4 per cent greater than 2 mm; whereas the original soil was 54 per cent less than 0·1 mm and 13 per cent greater than 2 mm.

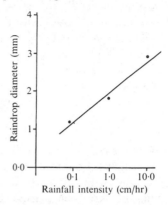

FIG. 6.6. The relationship between rainfall intensity and drop size. (From data in Laws and Parsons, 1943.)

Thus rainsplash and run-off preferentially transport smaller material, hardly surprisingly since a smaller force is required to dislodge it. But interestingly enough, rainsplash dislodged more material in the large-sized classes than did run-off. Hence rainsplash can dislodge more large material than run-off can carry away. Kirkby and Kirkby (1974) observed the upper limit for direct particle movement by raindrops to be about 50 mm, and particles 5 mm in diameter were often thrown up to 15 cm at a time. Laboratory experiments of rainsplash by Moseley (1973) showed that the quantity of soil caught on traps at varying distances away from a trough of sand subjected to rain declined exponentially with distance. The rate of decline was equal in all directions on level ground but, as slope angle increased, downslope transport and the maximum distance of travel increased. On a slope of about 25° over 95 per cent of the material travelled downslope. Another interesting point arising from Moseley's experiment is that *total* transport increased with slope angle by about six times from a slope of 0° to 25°, clearly demonstrating the importance of the downslope component of the impact force in particle movement.

Table 6.1

Raindrop size and momentum

Type of rain	Drop size (μ)	Maximum fall velocity (m/sec)	Momentum per drop (gm)
Drizzle or fine rain	200–500	1·5–3·0	6·2 × 10⁻⁴–0·019
Small raindrops	1000	4·0	0·21
Storm intensity raindrops	5000	9	58·9

Source: After data in Pettersen (1958).

Although rainsplash is capable of moving large numbers of particles quite long distances, its overall effect is sharply reduced by the run-off it generates. A layer of flowing water is a good cushion between raindrop impact and ground surface, and it has been observed that, as overland flow becomes operational over the whole slope, rainsplash becomes unimportant. Kirkby and Kirkby (1974) found this point to be reached at a flow depth of about 5 mm, but no doubt this depends on rainfall intensity and surface characteristics. Thus rainsplash is probably only significant in short storms where no run-off is generated, or close to a watershed where there is little catchment to generate flow. It is in this area that rainsplash is most often invoked as an erosive process of some importance in landscape development.

Overland flow

Once a measurable depth of flowing water has been generated on a slope, entrainment of particles as bedload or suspended load takes over from rainsplash as the dominant transport process. The hydraulics of fluid transport has already been discussed in section 6.2 and need not be elaborated on here, but a few points specific to overland flow should be noted. First, in a series of experiments on the hydraulics of overland flow (Emmett, 1970) it was found that flow was nearly always laminar, based upon the parameters of the flow, and yet quite significant sediment transport occurred. Whether this is conclusive evidence that laminar flow is capable of transport is to some extent debatable because, as Emmett noted, raindrop impact created some turbulence in the flowing layer. But the inference is that it is capable of transport.

Secondly, overland flow may occur as a continuous sheet across the slope, but it is more usual for concentration to take place along the flow direction. The preferred flow lines may be very indistinct, with barely recognizable banks, and may change position from storm to storm or even during a storm, but on occasion they may become quite well-defined channels or rills. It is intriguing to speculate on the reasons governing the organization of drainage on a slope. At the lowest level, explanations may be very simple. On a natural slope there will always be some irregularities tending to concentrate flow at some points; topographic hollows or the lee side of boulders, for example, provide converging flow patterns which tend to generate concentration. The wandering, anastamosing, flow patterns observed on unrilled slopes in semi-arid areas are largely related to this process. Water flowing down the trunks of trees from the canopy has also been observed to initiate rills (Segerstrom, 1950). But variations in surface topography or random concentrations of flow do not explain the development of well-ordered rill patterns, because rates of erosion and transport along rill-lines are much greater than on unchannelled sections of slope. Thus adjacent rilled and unrilled slopes would soon develop a discontinuity between them, with a corresponding increase in rates of erosion at the discontinuity progressively extending the rilled area to the whole slope. Emmett (1970) argued that whether a slope be rilled or non-rilled depended on the rate of erosion, since rills represent an organization of drainage capable of carrying more material which is ultimately related to the overall rate of removal from the base of slope. At dynamic equilibrium then, the rate of erosion on a slope is balanced by the rate at which material is carried from the base, and this ultimately determines the extent to which the slope is rilled.

Rate of removal by overland flow should be related to slope angle because flow velocities and bed stress tend to increase with slope, as well as the downslope component of particle weight. In experimental investigations a relationship with slope angle has been noted (Fig. 6.7) but it is much less

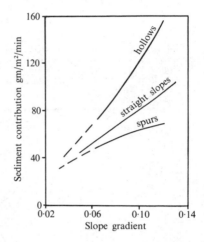

FIG. 6.7. The influence of slope on sediment yield in laboratory rainfall simulations. (After Moseley, 1974.)

easy to demonstrate on natural slopes. Reasons for this are the variability of vegetation cover (which is an important control in sheet erosion) on different slopes and also a tendency for transport rate to increase with slope length. The latter factor should not come as a surprise, since flow discharge will inevitably increase with downslope distance from the slope crest as catchment increases, and is especially noticeable in concentrated or rilled systems. The relationship is not so noticeable on the intervening unrilled areas, because the distance of flow to a rill may be roughly constant over the whole slope. Hence sediment transport depends on two topographic factors, slope angle and slope length, which are often correlated, so that assessing the importance of each is difficult.

'Available' sediment and sheet erosion

At any particular location one might expect the degree of surface lowering by sheet erosion to be fairly closely related to the amount of rainfall, or perhaps to maximum intensity of rainfall, in a period of time. Evidence from the laboratory and field, however, does not lend support to this intuitive relationship. This is largely because of the role played by weathering processes in releasing grains from the soil surface into a transportable state.

Even during laboratory experiments, where weathering is not expected to take place, the importance of some sort of surface preparation can be demonstrated. For example, Moseley (1974) and Emmett (1970) found declining sediment concentrations through time in water draining from experimental slopes under a constant rainfall intensity (Fig. 6.8a). This has been attributed to a process known as 'surface armouring', in which larger and therefore less transportable particles accumulate on the slope through time. Certainly there is a tendency for increasing grainsize through time, because

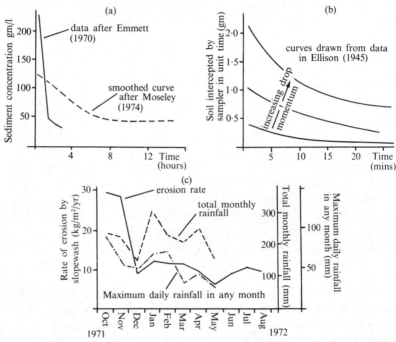

FIG. 6.8 The influence of surface preparation on sheet erosion. (a) From data in two articles; (b) From data in Ellison (1945); (c) Field observations from unvegetated gully-sides in the Black Mountain, Carmarthenshire.

rainsplash and overland flow selectively transport finer particles. This inevitably leads to declining transportability through time. However, results from Emmett (1970) tend to refute the armouring theory, since equally high rates of removal occurred when rainfall was resumed after a break as occurred at the initiation of the test. Emmett concluded that the process of wetting and drying was sufficient to loosen some grains by reducing cohesive forces between them, leaving a skin of easily removable sediment. A similar decline in rainsplash rates also occurs through a storm (Fig. 6.8b). The decline in this case has been attributed to surface sealing by inwashing of fine particles and compaction by rainbeat.

Weathering not only loosens material to cause high rates of sheet erosion at the beginning of each storm, it is also responsible for seasonal variations in rate. Rates of removal from gully sides in the Black Mountain

(Carmarthenshire) show high rates of removal in October and November, immediately following the summer period of lower run-off (Fig. 6.8c). Subsequent rates throughout the year are much lower, and show only a very vague relationship with monthly rainfall or maximum daily rainfall in any one month. During the summer, the surface is repeatedly wetted and dried, causing expansions and contractions of the surface to form a friable layer which is highly susceptible to removal in the early winter rains.

It is worth mentioning here that, theoretically at least, assessment of soil erodability is a problem in soil mechanics. In terms of chapter 4, the safety factor of a soil surface against erosion is equal to the ratio of soil strength (friction and cohesion) to bed stress imposed by the flow. Unfortunately, a strict mechanical analysis of stability is not possible, because factors such as vegetation cover and soil moisture are of overriding importance in determining surface resistance. Empirical attempts have been made to relate soil factors to erodibility. Most place emphasis on clay content, which is closely related to cohesion and moisture retention in the soil.

Weathering has wider implications in sheet erosion due to the part it plays in how slopes erode through medium to long time-spans. Some sort of concentration of flow on slopes is very common, and if the flow-lines remained constant through time they would tend to become an incised and therefore permanent channel system. If there is no net tendency for channels to become deeper through time, then some process must cause them to fill and alter course fairly regularly. This can take place during a storm, for example if a channel becomes diverted by local sediment deposition. But channel-fill can also take place seasonally with processes such as frost action, wetting and drying, animal disturbance and ploughing all providing mechanisms which tend to destroy rills. The rills need not necessarily change position with time as long as they do not progressively deepen. Ultimately, whether rills develop through time or are alternately cut and filled depends on the balance between weathering processes and surface lowering by sheet erosion. Again this is controlled by the rate of removal from the slope base. Permanent rill systems (section 6.4) can therefore be regarded as another organization of drainage adjusted to a higher rate of removal than seasonal rills.

Sheet erosion, vegetation and climate
The efficacy of all sheet erosion processes is limited by similar factors, which include soil factors such as cohesion and particle size, but probably of overriding importance is vegetation cover. Vegetation plays several roles in stabilization of soils to sheet erosion. Soil strength is increased by root binding and the soil is insulated from the direct effects of flowing water and raindrop impact by the vegetation itself and by the accumulation of litter and humus beneath it. Vegetation also significantly reduces the total amount of overland flow by promoting good soil structure and lower flow velocities, which favour greater infiltration. Finally of course, because velocity of flow is reduced, its capacity for transport is also lessened. Of all

these effects, probably the most significant is the reduction in total runoff due to improved soil structure. It is interesting to note that those factors, responsible for reducing overland flow and surface sediment loss, are the same as those increasing subsurface flow and therefore solutional loss. This relationship is reflected in the sediment to solute balance in stream transport (section 6.4).

On bare ground, annual or total storm transport by sheet erosion is dependent on total (annual or storm) rainfall, and rainfall intensity. Intensity provides the power or competence for splash erosion and to some extent for overland flow by increased turbulence, whereas total rainfall supplies the capacity for transport. In fact, transport rate tends to be a power function of total storm rainfall in the case of overland flow (Kirkby and Kirkby, 1974) and of total raindrop momentum in the case of rainsplash (Williams, 1969).

Different kinds of vegetation have their effect upon rates of transport by sheet erosion, and a few figures will serve to demonstrate this. For example, Kirkby and Kirkby (1974) quote comparative rates of erosion for cactus scrub, scrub oak and oak–pine forest as equal to, a quarter of, and one-twentieth of the rate for an immediately adjacent, bare ground area. These observations are from an area in Mexico with 600 to 900 mm rainfall per year. Ellison (1945 a and b) found that a cover of organic litter such as crop residue could significantly reduce soil loss in run-off, from 11·4 tonnes/hectare/year with 120 kg/hectare of cover to 0·4 tonnes/hectare/year with 816 kg/hectare of cover. Different crops were also shown to have varying effects on overland-flow velocity and rate of removal, with grass cover reducing erosion more significantly than wheat, which was in turn better than maize. Lowest rates of all are found on humid temperate, dense grassland slopes where rates become almost negligible.

The relationship between vegetation and sheet-erosion process implies a similar relationship between climate and sheet erosion, since natural vegetation cover is a function of climate. Generalizing climate solely in terms of mean annual rainfall, the relationship between plant cover and rainfall is shown by curve B in Fig. 6.9, with 100 per cent bare ground at no rainfall, and 100 per cent plant cover at a level of annual rainfall which will vary

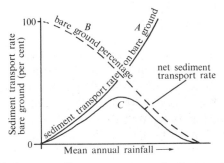

FIG. 6.9. The relationship between sediment transport rate by sheet erosion processes and rainfall.

according to other climatic factors, such as evaporation rate. Combining curves *A* and *B* gives curve *C*, the net transport rate curve, which shows maximum sheet erosion at moderate annual rainfall, or somewhere in the semi-arid to sub-humid climatic regime. The generalization made by Fig. 6.9 must be realized, since vegetation is only one factor of a number which combine to control transport rate. Not taken into account is the character of vegetation: for example whether it be canopy cover (forest) or ground cover, such as grassland. Rainfall intercepted by a tree canopy still has a long fall to gather momentum when it finally does get through. This may result in considerable rainsplash, especially as leaf drip frequently forms very large drops, and ground cover beneath trees is low. Also, trees concentrate flow down their stems and can cause channel initiation, mentioned previously. Finally, the modifications made to vegetation by man must be remembered. Man creates bare ground in high-rainfall areas by tillage and cropping, and damages vegetation cover by clearance and grazing. There can be little doubt that his activity has increased rates of sheet erosion by several orders of magnitude over large areas of the earth's surface.

6.4. Aqueous transport in permanent channel systems

Initiation of channel networks

From the previous section it follows that, if rills persist between storms, they increase their chance of survival, because subsequent flows will continue to follow and deepen them. Thus, a more organized state of drainage develops in which the simple, undifferentiated, overland flow is replaced by concentrated flow along a few specific lines.

Channel systems fall genetically into two distinct groups, though it must be stressed that continuous gradation exists between them. Systems in areas with very low infiltration-capacity soils, often semi-arid climates, flow only during storms and for a short time afterwards. Mostly they are dry, because little water is retained in the regolith. Naturally, therefore, most flow in storms is overland, and subsurface flow (throughflow) is negligible. In humid climates, where slopes are vegetated and infiltration capacities high, continuous flow occurs in river channels, though it is naturally highest during and after storms. Water which percolates into the ground during storms supplies river discharge in the dry periods. Semi-arid and humid drainage basins differ; for example, in the form of their slopes, a matter which is beyond the scope of this book, and in the length of cut channel per unit area (drainage density). Typically, semi-arid, ephemeral, channel networks have much higher drainage densities than humid river systems. Since water-flow is responsible for channel erosion, explanations of channel initiation and development should incorporate hydrologic factors. In fact, existing theories provide an excellent example of linkage between hydrologic and geomorphic processes. Initiation of permanent channels is discussed here only; development of networks is taken up in section 7.1.

Theories for the initiation of ephemeral channels or rill systems mainly stem from the work of R. E. Horton, who considered the relationship between slope hydrology and erosion (Horton, 1945). Horton's theory of channel-network development is based on overland flow as the main hydrologic process on new surfaces, and is illustrated in Fig. 6.10. Overland flow on a new slope will soon initiate a system of parallel rills by sheet erosion. However, since discharge increases downslope from zero at the slope crest, and since a critical discharge is necessary for sheet erosion to occur, there will be a belt of land adjacent to the slope crest which remains unrilled (Fig. 6.10a). Inevitably, a few rills will erode faster than the others and become dominant. A dominant rill will cut down more rapidly than its neighbours, and a local slope will develop towards it, leading to the ultimate destruction of neighbouring rills (Fig. 6.10b). This takes place by over-topping of rills at low points and lateral flow into the deeper rill, a process called *micropiracy* by Horton. Sub-parallel rilling begins anew on these new slopes facing the dominant rill, and in time micropiracy leads to the development of tributaries to the first dominant rill (Fig. 6.10c). This sequence of events continues, producing an integrated drainage pattern (Fig. 6.10d), whose limits are discussed in section 7.1. Support for Hortonian network development has been provided by many workers in low infiltration-capacity areas, of which only the work by Schumm (1956) will be mentioned here. Schumm studied channel development on a clay landfill and found close agreement with Horton's proposals, especially network geometry, development of network morphology through time, and the processes responsible for slope and channel erosion. Experimental work on small-scale laboratory slopes (for example by Flint, 1973, and Moseley, 1974) is also in agreement with Horton's sequence of events.

In the last fifteen years, Horton's proposals for channel development have been questioned with regard to humid areas, where overland flow is rare except in restricted localities. Bunting (1961, 1964), who was concerned with relationships between geomorphology and soil parameters, provided an alternative approach to channel evolution. Bunting found on some Pennine slopes that soil depth was far from uniform, and did not conform to the often-assumed trend towards increasing soil depth downslope. Instead he found linear zones where soil was up to twice the depth of the intervening areas. These so-called *percolines*, sometimes occupying topographic depressions but not always doing so, possess higher moisture contents at all times because they form lines of convergent subsurface flow and transmit more water than neighbouring areas. Percolines will develop in low points of the initial surface, and form a concentrated and partially integrated *subsurface flow system* (Fig. 6.11). These consistently wetter zones will be subject to faster rates of erosion by processes such as landslipping and solution and will deepen with time. In relating slope hydrology to erosion in humid areas Kirkby and Chorley (1967) recognized that a significant proportion of flood discharge of streams came from overland flow, but that it was derived from

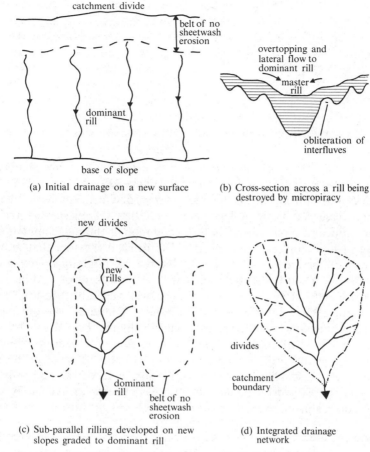

(a) Initial drainage on a new surface

(b) Cross-section across a rill being destroyed by micropiracy

(c) Sub-parallel rilling developed on new slopes graded to dominant rill

(d) Integrated drainage network

FIG. 6.10. Horton's approach to channel initiation and evolution.

only a small part of the drainage area. Infiltration capacities in British soils are normally in excess of rainfall intensities and so overland flow can occur only in areas of saturated soil: for example percolines, topographic hollows and areas adjacent to stream channels. The geomorphic significance of this is apparent. It is areas where overland flow occurs, notably topographic hollows and percolines, which provide the scope for development of channels, though subsurface pipes (chapter 3) may also provide a locus for channel initiation. Network development may be a slow process of progressive lowering along percolines and subsequently along permanent stream lines, but there is evidence to show that isolated storm events are significant in channel cutting, too. Large storms often cause intense local erosion, and frequently this is associated with extension of the drainage system, provided sufficient catchment exists to maintain the new channel. Hack and Goodlett (1960), for example, have shown this to be the case in the Appalachians.

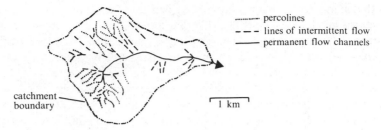

FIG. 6.11. Percoline distribution in a Derbyshire catchment. (After Bunting, 1964.)

Extension may be at channel heads or may cause entirely new channels to appear. Stream-head position is then, a balance between permanent flow, extension and cutting during intense storms, and continuous infilling by slope processes in between. Hack and Goodlett (1960) suggest that the smooth, concave, stream-head hollow, common in humid climates, is indicative of infill, and this is supported by their observation that grainsize tends to increase along channel heads; the reverse of along-stream sorting. This does not mean that there is currently a trend towards infilling. If cut takes place only in short-time events, one will inevitably see infill taking place in the longer, intervening time periods. A note of caution must be introduced here. Since infill does take place in channel-head areas, it is often difficult to say whether a percoline is a new, solutional feature or an old, infilled part of the surface channel network. Finally, it must be borne in mind that this discussion of channel initiation is very brief and only considers very simple hydrologic models. In areas of complex hydrology, channel initiation is proportionally more complex.

Sediment in river channels

The purpose of this section is to present a few examples of how sediment is derived in river channels, and how sediment concentration relates to discharge. The theoretical principles governing the initiation of sediment transport in fluids apply to flow in channels, but attempts to relate onset of erosion to bed-stress are restricted, due to the way sediment is derived. Only a small percentage is obtained from the river bed itself; the majority is already entrained in overland flow or is derived from bank scour and collapse. In the first case entrainment has taken place outside the channel, and in the second it is mainly due to material being thrown into suspension on collapse; in neither is it specifically related to stress at the base of the channel flow. Grainsize curves from parts of the river basin enable conclusions to be drawn about sediment supply, and results differ between basins. To take just two examples: Rendon-Herrero (1974) found bed and bank material in a small Pennsylvanian watershed to be much coarser than suspended load, and concluded that most sediment was supplied in overland flow. On the other hand Carson *et al.* (1973) found a close association between bank material and suspended load in a Canadian catchment, and were able to show that bank scour and collapse provided most sediment. Clearly, sediment source is related to hydrology, in that if overland flow occurs it is likely to supply sediment. Carson *et al.* (1973) also found that 70 per cent of the sediment load came from the main channel, which, however, constituted only 18 per cent of the total network. This rather surprising result suggests localized erosion on the main channel, but in fact it may be explained by the fact that sediment load was found to be a power function of total run-off. Since run-off increases downstream, the main channel would be expected to possess a greater erosive capacity.

Bedload in streams is difficult to measure, and all estimates of its importance must be treated with uncertainty. Certainly bedload movement relies upon the grainsize of available materials and assumes greatest importance when sediment supplied to streams is in the coarse-sand to gravel range and up to cobble or boulder size. Finer sediment will move as suspended load in high flows and, while it may constitute bedload in lower flows, it will move only slowly and will not be a significant part of total load. Estimates of bedload vary from 0 per cent to 50 per cent or more of the total sediment load, but it is probably normally from 5 to 20 per cent. Because of the difficulties in measuring bedload, the ensuing discussion is in terms of suspended load only, though much will be applicable to bedload. From theoretical considerations, sediment load increases with water discharge, and there should therefore be a relationship between load and discharge at a point on a stream. In practice, this relationship is established empirically for a river and is called the sediment rating curve. It has the general form:

$$T = Qc = KQ^n,$$

where T is transport rate, Q is the discharge, c is the mean sediment concentration and K and n are constants. Throughout a storm, sediment load (concentration and transport rate) shows an increase relative to water discharge, an example of which is shown in Fig. 6.12a. The sediment peak may coincide in time with the water peak, as in the example illustrated, or it may be earlier. Sediment concentration and transport rate tend to be higher throughout the rise in discharge through a storm than at equivalent discharges when river level is falling (Fig. 6.12b). Because of this effect, sediment concentration throughout a storm shows a hysteresis or 'looped' relationship with discharge (Fig. 6.12c). This should be borne in mind when using a rating curve to predict sediment concentration. There are several factors which may explain the hysteresis effect. It was mentioned in section 6.3 that the highest sediment yield in overland flow occurred soon after the start of rainfall, with declining rates subsequently. Thus one would expect most sediment in overland flow to reach the channel during rising discharge. There is also the obvious point that it has generally stopped raining by the time discharge is falling, and so there will be little overland flow anyway. Sediment derivation in the channel may also be greater during rising discharge. Obviously, the most easily erodible sections of bank will collapse

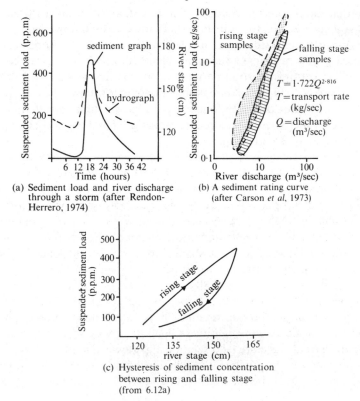

(a) Sediment load and river discharge through a storm (after Rendon-Herrero, 1974)

$T = 1·722Q^{2·816}$
T = transport rate (kg/sec)
Q = discharge (m³/sec)

(b) A sediment rating curve (after Carson *et al*, 1973)

(c) Hysteresis of sediment concentration between rising and falling stage (from 6.12a)

FIG. 6.12. The relationship between sediment load and discharge in rivers.

during rising discharge, leaving more resistant material exposed as flow recedes. Another factor of possible importance is that water slope is greatest throughout rising discharge. This is because maximum water level at flood peak is upstream on the rising limb and therefore adding to the channel slope, but during falling discharge the peak is downstream and therefore subtracting from the slope. Since velocity and therefore bed stress increase with water slope, the capacity for transport will be lower during falling discharge.

Because sediment transport depends on a critical velocity or discharge of flow being exceeded, and because that threshold is related to grainsize, some changes in sediment properties along channels are to be expected. Bed material reflects these changes, though with the proviso that it may not have been deposited only in recent flows but may remain from very old flows as well. Thus, very large particles may survive in channels for long periods and may not be related to recent hydrological conditions at all. Two processes tend to maintain progressive variation in sediment quality along channels. First, rivers sort sediment by transporting material below a threshold grainsize and leaving coarser and better-sorted material behind. Consequently, predominantly finer material reaches the lower part of the system. Secondly, grain breakage and wear take place with transport, also resulting in a declining grainsize downstream. In many channel systems these downstream changes are masked by the continuous supply of sediment of mixed grainsize from the channel banks, but it is possible to appreciate them in systems with a discrete input at the channel head and limited bank supply. An example of such a stream is one which issues from the base of a glacier or ice sheet, and data from one Icelandic glacial stream is presented in Fig. 6.13. In particular, note that sorting improves downstream due to

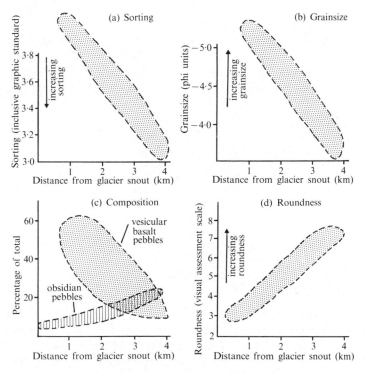

FIG. 6.13. Changing bed sediment properties along a stream draining from a glacier. (After Huddart, unpublished data.)

selectivity of flow (Fig. 6.13a) and that mean pebble size progressively declines (Fig. 6.13b). How much the latter is attributable to breakage as opposed to selection over a short length of channel is difficult to answer, but unequivocal proof that grain breakage does take place is gained from the relative persistence of different rocktypes downstream (Fig. 6.13c). Obsidian fragments, only 5 per cent of the bed material near to the glacier, increase to 20 per cent over 3·5 km of channel, while less resistant vesicular basalt declines from over 50 per cent to about 10 per cent in the same distance. Furthermore, particle roundness increases downstream, showing the effects of wear on the pebbles (Fig. 6.13d). Changing grainsize distribution over longer travel distances is taken up in section 7.4.

Solutes in Rivers

Solutes in rivers are derived predominantly from some form of subsurface flow, either seepage or pipe-flow (chapter 3), because contact between water and soluble materials is longer and better within the soil. Almost 90 per cent of the dissolved load of rivers is made up of just four ions (HCO_3^-, SO_4^{--}, Ca^{++}, SiO_2) with a number of minor constituents comprising the remainder (Table 6.2). In some rather uncommon situations specific constituents may assume greater or lesser importance than Table 6.2 suggests, depending upon their availability within the river basin. Discussion of the behaviour of solutes in streams is confined here to two constituents, calcium and silica, and to total load, since these demonstrate the principles adequately. Unlike sediment, solute concentration tends to *decline* with increasing discharge. This inverse relationship causes solute concentration to decline through a single storm, well exemplified by Fig. 6.14a, which shows calcium carbonate

Table 6.2

Mean composition of dissolved solids in river water

ion	% of mean total concentration
HCO_3^-	54·6
SO_4^{--}	9·6
NO_3^-	0·9
Ca^{++}	13·9
Mg^{++}	3·4
Na^+	2·5
K^+	2·1
Fe^{+++}	0·7
SiO_2	12·3

content of a stream rising from limestone in the Mendips over a storm period. The decrease may be attributed to dilution by near-surface flow during storms, which does not possess such a high solute concentration. A hysteresis effect is again noticeable, since rising discharge samples tend to have lower ionic concentrations than falling discharge (Fig. 6.14b). Total dissolved load follows a similar pattern, demonstrated by the rating curve of solute concentration against discharge for a large river, such as the Mekong (Fig. 6.14c). The rating curve shows an inverse relationship between solute concentration and discharge of general form:

$$I = KQ^{-n},$$

where I is the total ionic concentration, Q is the discharge and K and n are constants. Again hysteresis is often noted in the relationship between solute load and discharge.

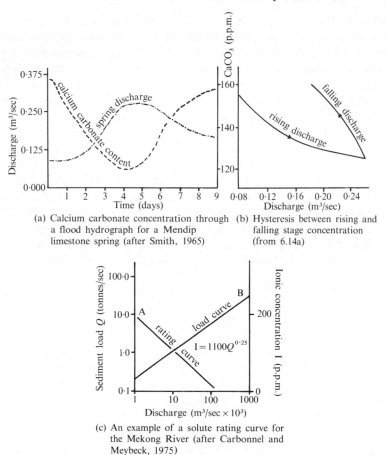

(a) Calcium carbonate concentration through a flood hydrograph for a Mendip limestone spring (after Smith, 1965)

(b) Hysteresis between rising and falling stage concentration (from 6.14a)

(c) An example of a solute rating curve for the Mekong River (after Carbonnel and Meybeck, 1975)

FIG. 6.14. The relationship between dissolved load and river discharge.

The one ion which seems to be exceptional in its relationship with discharge is silica, since its concentration remains nearly constant through storms (Davis, 1964). This has been attributed to several mechanisms, for example: solution from the larger surface area of suspended sediment in streams during floods; solution of distorted silicate lattices resulting from grain impact in the stream; or possibly silica extraction from stream water by diatoms during low flows. None of these is entirely satisfactory, since they do not account for silica concentration being maintained through flood-peak dilution, even though rates of silica uptake from almost all silicate minerals have been shown to be very slow.

As pointed out by Douglas (1964) the importance in erosional terms of these variations in solute concentration with discharge is questionable. Ionic concentration varies inversely with a power function of discharge (see the rating curve form above) and for the example of the Mekong (Fig. 6.14c), this means that a tenfold increase in discharge causes only a halving of the ionic concentration. In many rivers the effect is much less than this. Consequently, solute *load*, which is the product of discharge and concentration, still continues to increase with discharge in a way similar to sediment load (curve *B*, Fig. 6.14c). The fact that relationships between transport loads and discharge tend to be power functions has an important bearing upon the relative importance of large but rare floods and more normal discharges in over-all erosive loss (section 7.3).

The balance between solute and sediment load in streams

Average world erosion figures show rates of erosion of about 97 tonnes/km²/year and 37 tonnes/km²/year due to sediment and solute loss respectively. The figures vary according to methods of calculation, but on average solutes comprise about 38 per cent of the total load of the world's rivers. However, Tables 6.3 and 6.4 show that the balance is far from constant from one river to another. In general, for example, Asian rivers have

Table 6.3

Sediment load of some major world rivers[a]

River	Continent	Mean Discharge $(m^3 \times 10^6)$ per year	Mean Sediment Load (tonnes/yr) ($\times$ 1000)	Mean Sediment Loss (tonnes/km²/yr)	Surface Lowering Rate (mm/1000 yrs)	Mean Sediment Concentration (kg/m³)
Huang	Asia	47 300	2 040 000	2 900	1 160	4.3×10^{-2}
Colorado	N. America	4 900	146 000	430	172	3.0×10^{-2}
Irrawaddy	Asia	428 000	324 000	930	372	7.7×10^{-3}
Ganges	Asia	371 000	1 570 000	1 600	640	4.3×10^{-3}
Missouri	N. America	61 600	236 000	179	72	3.8×10^{-3}
Indus	Asia	175 000	470 000	520	208	2.7×10^{-3}
Nile	Africa	89 300	120 000	40	16	1.3×10^{-3}
Yangtze	Asia	688 000	540 000	560	224	7.9×10^{-4}
Mississippi	N. America	560 000	338 000	110	44	6.1×10^{-4}
Mekong	Asia	349 000	184 000	490	196	5.3×10^{-4}
Orinoco	S. America	715 000	93 000	103	41	1.3×10^{-4}
Danube	Europe	195 000	21 000	27	11	1.0×10^{-4}
Volga	Europe	253 000	20 000	16	6	8.2×10^{-5}
Amazon	S. America	5 720 000	390 000	68	27	6.9×10^{-5}
Congo	Africa	1 250 000	70 000	18	7	5.7×10^{-5}
Ob	N. Asia	394 000	15 400	7	3	3.9×10^{-5}
Niger	Africa	192 000	4 870	5	2	2.5×10^{-5}
Yenisey	N. Asia	549 000	11 400	5	2	2.1×10^{-5}
Rhine	Europe	69 700	495	4	2	7.2×10^{-6}

[a] Ranked by sediment concentration (from data in Holeman (1968)).

Table 6.4

Percentage of total load carried in solution for some large rivers

	%	
Colorado	12	Below average
Mekong	25 ⎫	
Amazon	32 ⎬	About average
Mississippi	36 ⎭	
Congo	76	Above average

higher than average sediment concentrations, and over 80 per cent of the world's river sediment is derived from that continent. This is partly a reflection of its size, but even when this is taken into account, sediment loss per unit area of land in Asia is four times greater than the world average. Other rivers can have very high sediment loads; a specific example is the Colorado (U.S.A.) at the Grand Canyon.

Clearly, those factors which influence the balance between surface and subsurface water-flow will be those controlling the sediment-to-solute balance. The most obvious of these is the percentage of vegetation-free land in a river basin, which was shown to be broadly a function of climate (section 6.3). Thus, just as the peak for rainsplash and slopewash erosion falls in the semi-arid to sub-humid climatic zone (Fig. 6.9), so one may expect highest sediment concentrations in those same areas. Using mean annual runoff, expressed as a depth of water over the land surface, as a crude indicator of climate (it is the net amount of water in a basin after rainfall and evaporation have been considered) data from 168 United States gauging stations (Fig. 6.15) show the influence of climate very well. Solute concentration declines with increasing annual runoff, for reasons already discussed, but the total *load* increases because the relationship between concentration and runoff is not linear. When solute and sediment data are compared for American rivers (Fig. 6.15) it is apparent that the percentage of material carried in solution increases steadily, up to nearly 50 per cent at mean annual runoff of 500 mm. Notice that sediment load in the U.S.A. peaks at about 10 mm annual runoff (semi-arid conditions) and slowly declines beyond that level.

The importance of a second factor in the sediment-to-solute ratio, namely relief, can be demonstrated by erosional loss from different parts of a large river basin, such as the Amazon (Gibbs, 1967). Areas of higher relief should have greater sediment loss, because many processes of sediment derivation (sheetwash, landslides) act faster on steeper slopes. Amazon data show this trend (Fig. 6.16) but also show that solute load increases in areas of higher relief, probably reflecting faster discharge rates through the regolith on steeper slopes. However, since it is clear (from Fig. 6.16) that sediment loss

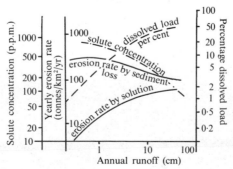

Fig. 6.15. Solute and sediment load as a function of runoff for U.S.A. rivers. (After Langbein and Dawdy, 1964 and Langbein and Schumm, 1958; combination of two diagrams.)

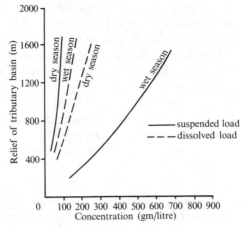

FIG. 6.16. The influence of basin relief on dissolved and suspended load in the Amazon Basin. (After Gibbs, 1967; combination of two diagrams.)

increases much more rapidly with relief than solute loss, sediment loss assumes greater over-all importance in the more mountainous parts of the basin. Table 6.5, a summary of erosional losses from sub-basins within the Amazon basin, shows that the percentage of dissolved material is less than the world average in mountainous and mixed catchments, and greater in tropical catchments. However, the range is much greater in tropical catch-

Table 6.5

Solute to sediment balance and erosion rates in parts of the Amazon Basin

Name of tributary basin		Rate of erosion (tonnes/km²/yr)	% of total load in solution
Amazon at mouth		116	32
Ucayli	mountain	459	33
Maranõn	catchments	344	27
Napo		213	14
Madeira	mixed	199	27
Ica	catchments	79	22
Japura		23	48
Juruá		83	40
Javari		80	14
Purus		73	41
Jutaí		46	11
Negro	tropical	20	50
Araguari	catchments	17	59
Coari		15	87
Tefé		15	85
Tapajós		5	76
Xingu		4	75

Source: After Gibbs (1967).

ments again, showing the effect of relief. The predominance of chemical loss in the lower parts of the Amazon is also shown by a contrast in sediment quality between the mountainous and tropical areas. Mountainous areas are dominated by unaltered mineral detritus, whereas sediment from the tropical areas consists largely of clay minerals, which are secondary products of chemical weathering (Gibbs, 1967).

Though the importance of climatic and topographic factors in the sediment/solute balance cannot be denied, the effect of human activity must not be forgotten. Man both increases sediment load by agricultural practice on the one hand, and decreases it by building dams which trap the sediment on the other. Solutes are also affected, for example, by industrial effluent and by dissolved fertilizers in surface run-off. Consequently, the measured load of a river system may in many cases be a reflection of human interference rather than of natural erosion processes.

6.5. Transport of sediment in wind

The mechanics of wind transport have already been discussed in section 6.1, where general principles of fluid transport were considered. Air movement is in close agreement with general fluid mechanics; for example, the logarithmic velocity profile predicted from the theory of fluid flow is well established for air movement. Determination of the rate of sediment transport in air has also been dealt with in section 6.1 and so there is no further need to dwell upon the mechanics of air transport here.

Where movement of sediment in air differs from movement in water is in the relative proportions of the sediment load in bedload, saltating load and suspension. The amount in suspension is closely controlled by the grainsize distribution of the material being eroded. This is very variable, but in general materials subjected to wind erosion fall in the sand grainsize range. The two most important wind sediment transport environments are deserts and beaches. A considerable amount of research in North Africa, the Near East and Australia has demonstrated that the most significant sources of wind-blown detritus are areas of extensive fluvial sediments which are deposited after infrequent but intense storms. In some cases they may be fossil deposits from periods of greater runoff. This material is usually of sand size, and so sand-size material predominates in the wind system as well. In short, wind can only transport what is available. The normal range of wind velocities is only capable of transporting silt- or clay-size material in suspension, and so suspension is a small part of total sediment transport in most wind systems. There are exceptions however. Glaciers produce a considerable amount of silt-size material by grinding and scouring of bedrock and boulder surfaces, and if this is deposited near to the glacier front it becomes highly susceptible to wind suspension due to sparse vegetation cover. Large quantities of airborne dust were derived in this way during the Pleistocene, and accumulated as a deposit known as loess over much of Northern Europe and Asia. Where soils have been devegetated, either naturally by fire or artificially by agriculture, there is also a potential for appreciable suspended sediment transport by wind.

As a broad generalization however, grainsize and wind velocity in areas of appreciable wind transport favour saltation. Estimates vary, but on average it seems that about 75 per cent of all sediment movement is by saltation, in the desert zones at least. The remaining transport is by surface creep, which is a combination of rolling, sliding and pushing by impacting, saltating particles. In contrast, water transports little material by saltation. Because of the greater density and viscosity of water, saltating particles experience much greater resistance, and so fewer particles saltate and they bounce less far.

Surface factors influencing wind transport

The mechanics of sediment movement discussed in section 6.1 relate to movement by wind over a mobile bed of cohesionless, dry material. Sand transport is naturally much less where grain attractions exist. The most important factors in this respect are vegetation cover and surface moisture, though surface roughness and grain aggregation also play an important role.

Vegetation insulates the ground from direct wind action in the same way as from direct water action. Thus a very dense vegetation mat with 100 per cent cover will preclude almost all effective wind transport. But even an open cover reduces transport, largely by changing the wind velocity profile near to the ground. Vegetation determines the height of the *aerodynamic surface*, above which air flow is unimpeded and assumes its normal logarithmic velocity profile. At the aerodynamic surface, wind velocity may be effectively zero with very dense cover which is impermeable to air flow, or it may be much reduced if some air flow occurs within the vegetation layer. In any event, wind velocity and velocity gradient and therefore *shear stress* are much reduced at the ground surface, reducing the likelihood of sediment entrainment. In effect, the vegetation absorbs part or all of the shear stress imposed by the wind. Fig. 6.17 shows velocity profiles under similar wind conditions for two vegetation covers; growing wheat whose height is 5 cm in still air and millet stubble of maximum height 53 cm. The low, permeable wheat cover shows much higher velocity gradients close to the ground than the stubble, although some air movement was detectable within the latter (shown by the dotted line in Fig. 6.17). Vegetation cover is often taken into consideration as an empirical constant in sand-transport formulae.

Interstitial water results in particle attraction due to surface-tension forces in the water layers at grain contacts. Thus it is to be expected that transport rates should be reduced with increasing moisture content in the surface layer. Fig. 6.18 is a plot of sediment-transport rate against equivalent moisture content at three different wind speeds, blowing over the same silty soil surface. Equivalent moisture content (m_e) is defined as:

$$m_e = m/m_{15a}$$

where m is the soil moisture and m_{15a} is the moisture content of the soil held at 15 atmospheres tension. The significance of the 15 atmosphere-tension

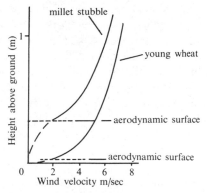

FIG. 6.17. Wind velocity profiles over vegetated surfaces. (After Chepil and Woodruff, 1963.)

level is that it is the maximum suction that plants can exert in extracting water from soil. Thus at lesser moisture contents plants wilt. m_{15a} is a function of soil grainsize, because smaller grains can hold more water at a given tension than larger grains. In sands it is about 4 to 6 per cent, while it is nearer 10 to 15 per cent in silts and 20 per cent in clays. Transport rates increase rapidly with declining equivalent moisture until they are nearly constant above $m_e = 0.2$, which is the level at which soil is effectively dry. (This corresponds to a tension of about $30atm$ in the example quoted, which is the tension at which no more water is extractable by evaporation.) Another way of looking at the influence of moisture content is that it affects the critical or threshold value of bed stress at which transport just begins (Svasek and Terwindt, 1974). Actually, two factors are involved in the influence of moisture content on wind transport: first, surface tension, which increases initially with moisture content but then declines at high values; and secondly,

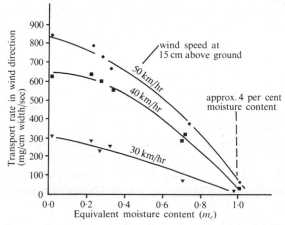

FIG. 6.18. The influence of soil moisture on wind transport over a sand surface. (Plotted from data in Chepil and Woodruff, 1963.)

reduction in exposed grain area to the wind which is greatest at saturation. Thus transport rate continues to decline with increasing moisture content even though surface tension is reduced at high moisture values, because the grains are protected from the wind by a film of water. Since rates are low above 15*atm* tension, however, this is only of academic interest. The influence of moisture on sand movement was shown very well in a study of sand dunes in County Mayo, Eire, by Crofts (1971). The predominant westerly winds in this area are usually rain-bearing and consequently the surface is wet when these winds blow. In fact, net sand transport is from the south-east, a response to rarer but drier winds.

Other surface factors affecting wind transport include grain aggregation and ground surface roughness. If grains aggregate, then higher wind speeds are necessary to move the aggregates. Aggregation is favoured by fine-grained soils and soils with appreciable clay content. Surface roughness increases the height of the aerodynamic surface and so decreases the transport rate. These factors are discussed in more detail by Chepil and Woodruff (1963).

Wind transport systems

Fluid must flow along the direction of applied stress. In water this is simply the line of steepest slope on the land surface, and leads to concentration of flow along specific lines and progressive integration of the flow system into a network (section 6.4). In air no such tendency to concentrate in the direction of flow exists, and consequently it is not possible to define a system which flows to a single exit through which all fluid and sediment must pass. Air flows along pressure gradients which vary over many different time-scales. But at many places there is a strong tendency for wind to blow predominantly from a particular quadrant, and so there will be a net sediment-transport direction. One can therefore define a transport system on the basis of sediment-flow paths. Figure 6.19 shows just one example of a sandflow system from the Sahara desert in North Africa (Wilson, 1971), based on windflow data. Maps may also be produced from the orientation of sand bodies such as dunes, but some sand bodies are very large and take a long time to accumulate. Thus they will take a long time to adjust to a new wind system, and may reflect past rather than present conditions. On the other hand, they give an extension of transport-direction data much further back than wind records, which usually extend back comparatively few years. Returning to the sandflow map (Fig. 6.19), it may be noticed that over this large area a very well-defined and organized transport system can be discerned. That the pattern is not chaotic is to be expected, since particular wind directions predominate in this region. There is a well-defined divide, north of which wind blows to the Mediterranean, and south of which it swings round towards the Atlantic. In fact, on a very large scale, the Sahara is not unlike two drainage basins, since there are well-defined areas yielding sediment in separate directions. The differences are: that at any given time sediment may

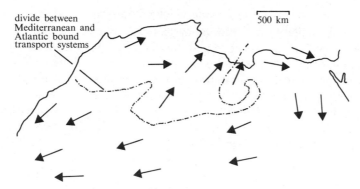

FIG. 6.19. Net sandflow directions in the Sahara. (After Wilson, 1971.)

move in any direction, and that the flow paths do not converge and join but remain more or less parallel.

In a wind-transport system such as the Sahara, sediment is stored in much the same way as in a river drainage basin. Thus sediment is dropped from the flow when wind velocities fall below a critical level, and remains in store until the next event. There is also a tendency for net accumulation in areas where the wind becomes overloaded with sediment. This usually occurs in areas where flow lines are convergent or where velocity is reduced. Topographic hollows or the lee side of large topographic features often supply such conditions. A sand accumulation will continue to grow until inflow of sediment is balanced by outflow, which will occur only when wind conditions in the accumulation zone have changed in such a way that the effects of the original irregularity are counteracted.

6.6. Transport in glaciers

Accumlations of ice in the form of continental ice sheets and valley glaciers are now restricted mainly to polar regions and high mountains. It is as well, however, to remember that, while ice is now of rather small importance in terms of world sediment transport, in the past it has played a major role. Indeed, during the Pleistocene a considerable percentage of the world's land surface was under ice at some time, a fact which is demonstrated by large accumulations of glacially derived sediment over much of the northern hemisphere's lowlands and erosional landforms of glacial origin in much of its highlands. Accumulation of ice in sufficient quantities to form glaciers is of course climatically controlled. However, it may not be due simply to a reduction in mean annual temperature; other climatic factors such as winter precipitation or mean annual cloudiness may be involved. A discussion of the climatology of glacier accumulation is beyond the scope of this book, and this section is confined to the mechanics of ice deformation and glacier flow, and to the way in which sediment is derived, transported and deposited in a valley glacier system. The purpose is not to cover glacial transport in detail but to highlight the contrast with aqueous transport.

The mechanical properties of ice and glacier movement

A superficial similarity between valley glaciers and rivers led early workers to draw an analogy between ice and fluid of high viscosity. While there are some similarities between glacier movement and fluid behaviour, particularly in that velocity of flow declines with depth and towards the valley-sides, recent experimental work has demonstrated that ice is not an ideal fluid.

Uniaxial compression and extension tests on single ice crystals have shown that ice has no yield stress but deforms under any level of applied stress. Thus far, it shows fluid properties. The *rate* of strain ($\dot{\varepsilon}$) is related to stress (τ) by a simple relationship:

$$\dot{\varepsilon} = A\tau^n, \tag{6.7}$$

where A is a temperature-dependent constant, analogous to fluid viscosity, and n is another constant. Whereas in a fluid n would equal one, in ice it is from 1·9 to 4·5. In other words, the relationship between strain rate and applied stress in fluids is linear but in ice it is a power curve (Fig. 6.20). The value of n is also variable, increasing with confining pressure. Furthermore, experiments on single ice crystals show them to be anisotropic with respect to applied stress. The crystal structure of ice consists of hexagonally ar-ranged water molecules stacked together in parallel sheets. Deformation is preferential, parallel to these sheets, with very much higher levels of stress being necessary to cause an equivalent rate of deformation oblique to the structure, which is not typical of a fluid. Polycrystalline ice is more isotropic on a large scale, because crystals lie at all inclinations within the mass. Higher levels of stress are required to deform polycrystalline ice than to deform single crystals, due to interactions between the grains, though at very high stresses recrystallization of grains parallel to the direction of applied stress increases the rate of deformation.

In fact, stresses less than about 100 kN/m² cause small deformations in ice, while those above eventually cause very large displacements. Hence, it is justifiable to approximate ice deformation to ideal *plastic* behaviour, with a yield strength of about 100 kN/m² (Fig. 6.20). Of course, ice is neither ideally plastic nor viscous but lies somewhere between the two, and is described by the flow law in equation (6.7). What has been said above applies strictly to

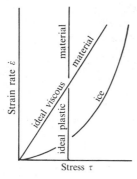

FIG. 6.20. Relationship between strain rate and applied stress in ice.

pure ice; incorporated debris changes the mechanical behaviour of the mixture. As one would expect, as debris is added to ice, the resultant material becomes progressively more similar to a cohesionless aggregate in its response to stress. Since there is often a concentration of debris at the sole of a glacier one might expect the ice to become more plastic in this region, enhancing the possibility of basal sliding.

Assuming ice to be a perfect plastic material, the justification for which was mentioned above, one can calculate the basal shear stress (τ_b) of a parallel-sided ice sheet of thickness h from:

$$\tau_b = \gamma_i h \sin \alpha$$

where γ_i is the unit weight of ice and α is the surface slope of the sheet. When τ_b is equal to the assumed plastic yield stress for ice (τ_c), sliding would begin at the glacier sole and so ice thickness and basal shear stress could not increase further. From this it follows that the thickness of ice can be calculated as a function of surface slope viz.

$$h = \tau_c/\gamma_i \sin \alpha.$$

Although this derivation is only strictly valid for an ice sheet, correction factors allow it to be applied to the valley glacier situation. Plasticity thus implies that $h \sin \alpha$ is constant and, in support of this, thinning and extension occur where glaciers flow over steep sections of bed whereas thickening and compression occur where glaciers flow on to lower slope angles. This leads to the well-known pattern of crevasses and thrusts seen in *extension* and *compression* zones respectively in glaciers (Fig. 6.21). Also worth noting in defence of the 'plastic approximation' to glacier movement is the fact that basal shear stress beneath ice sheets usually falls in the range of 50 to 150 kN/m², a tolerably good approximation to the assumed plastic yield stress of ice of 100 kN/m².

But glacier flow differs appreciably from plastic flow with respect to the distribution of velocity within the ice. Plastic sliding at the glacier bed can constitute quite a large amount of the total movement of some glaciers, but in most there is a general decline in velocity with depth, with a greater or lesser amount of basal sliding (Fig. 6.22). From the flow law for ice, and assuming that *simple shear* takes place (simple shear is the progressive glid-

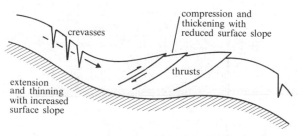

FIG. 6.21. Extension and compression in glacier flow.

ing of one layer over the next, rather like a pack of cards being pushed), then the rate of change of velocity (u) with depth in the ice at any depth y is given by:

$$du/dy = 2A[\gamma_i(h - y) \sin \alpha]^n \quad \text{(Paterson, 1969.)}$$

Hence, the slope of the velocity profile at any point within the ice is proportional to a power of the product of depth and slope, the latter being constant at any vertical section. Thus, from the flow law, the velocity of flow is a power function of depth, with zero basal velocity, which is not too dissimilar from observed velocity profiles from real glaciers (Fig. 6.22). An interesting point from the velocity profile derived from the flow law is that, since velocity is a function of surface slope but not bed slope, uphill flow is theoretically possible. Observations made in once glaciated valleys indicate that uphill flow is not uncommon.

We have already seen that some basal sliding takes place in glaciers, even though velocity profiles derived from the flow law preclude it. The importance of basal sliding is however very variable, from about 10 per cent to 90 per cent of total movement with a mean around 50 per cent. There appears to be a distinction between glaciers which are below their pressure melting-point throughout and those which are at their pressure melting-point in this respect. The former, cold-base glaciers have a frozen contact with underlying bedrock, and since frozen debris layers or frozen ice-to-rock contacts are much stronger than pure ice, basal sliding is precluded. On the other hand, temperate glaciers are not so constrained and are free to slide at their bases. This distinction between cold-base and temperate glaciers is an important one and will be returned to below when ice transport is considered. It appears that two mechanisms contribute to the sliding of a glacier past obstacles on its bed. First, the pressure gradient between the upslope and downslope sides of small obstructions causes pressure-melting on the up-

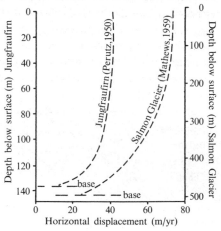

FIG. 6.22. Velocity profiles through two glaciers.

stream face, and water flow to the downstream face, where it refreezes. Flow of latent heat from the refreezing face to the melting face of the obstacle is intrinsic in this mechanism, limiting it to small basal irregularities. Secondly, obstacles reduce the available cross-section through which the ice can flow at its base, thus increasing the rate of plastic flow around these obstructions. This mechanism seems to be more applicable to larger irregularities. Other factors, such as water pressure in cavities at the glacier sole, or simply reduced resistance to sliding due to air-filled cavities in the lee of obstructions, may also contribute to basal sliding.

Sediment incorporation and transport in glaciers

Controversy exists among glaciologists concerning the ability of ice to erode intact rock. While this controversy is beyond the scope of this book, it is worth noting that pure ice cannot cause direct erosion of intact rock simply because its shear strength is considerably less than almost all fresh rock. Nevertheless, this does not exclude abrasion between debris incorporated in the base of the ice and underlying bedrock, nor does it exclude direct detachment of weathered or loosened rock fragments by the ice. Refreezing of ice in the lee of bedrock obstructions may, for example, lead to loose joint-blocks being plucked away as the ice moves on.

Sediment is transported in glacier systems on the ice-surface (supraglacial), within the ice (englacial) and at its base (subglacial). Much of the debris transported is derived not from subglacial bedrock erosion but from sub-aerial processes acting upon the valley sides above the glacier. Debris thus accumulates supraglacially, though much is subsequently transferred into englacial and subglacial positions by the addition of more ice above it and by internal movements in the glacier. A good illustration of how these processes operate is gained from the internal distribution of debris in a small Norwegian cirque glacier, reported by McCall (1960). An annual layer of supraglacial debris is added by rockfalls in spring and summer to the glacier, and is subsequently buried beneath a fresh layer of ice throughout the winter season. Thus, the debris is preserved englacially, though because of the way the glacier moves some of the annual layer is transferred to a subglacial position (Fig. 6.23). Most of the movement is by basal sliding on its concave base, maintaining contact between debris layers and the base, and pressure melting of basal ice tends to release debris to the sole. In the lower part of the glacier, net summer melting re-exposes the annual debris layers, once again transferring material to the surface.

A comprehensive treatment of glacial transport is not possible here owing to limitations of space; consequently, reference is largely confined to a series of papers by Boulton (1967, 1970a, 1970b, 1972a, 1972b), whose work on valley glaciers in Spitzbergen illustrates the properties of a glacial transport system rather well. In particular, we shall see the contrast between the ways in which debris is incorporated, transported and deposited in glaciers as opposed to ideal fluid systems. As in many glaciers, those in Spitzbergen

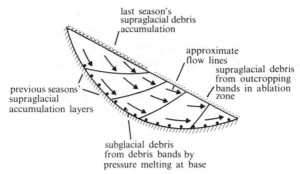

last season's
supraglacial debris
accumulation

approximate
flow lines

supraglacial debris
from outcropping
bands in ablation
zone

previous seasons'
supraglacial
accumulation layers

subglacial debris
from debris bands by
pressure melting at base

FIG. 6.23. Debris incorporation and flow in a cirque glacier. (McCall, 1960.)

derive much of their sediment load from sub-aerial rather than subglacial processes, and this sediment is ultimately transferred to the glacier's interior and sole by accumulation of more ice and internal glacier movements. The amount of sediment supplied to glaciers depends on the erodibility of the country rock. For example, glacier tongues in Icelandic valleys, where the country rock is mainly soft ashes and easily fragmented lavas, tend to be quite 'dirty' in comparison with many Norwegian glaciers, which pass through areas of very tough, metamorphosed sediments and igneous rocks.

Boulton stresses the important distinction between cold-base and temperate glaciers in debris-incorporation and transport. The cold-base glaciers in Spitzbergen develop a basal zone of sub-glacial and englacial debris which is many metres thick. This basal debris layer is itself divided into many thin layers of debris-laden ice (a few millimetres thick) separated from each other by clearer ice bands. The bands show a systematic downstream incorporation of material from top to bottom of the debris zone (Fig. 6.24). Two possibilities have been suggested to explain the development of a thick debris layer in cold-base glaciers. The first is that the layers represent thrust planes in the ice which have transferred sub-glacial debris upwards. An argument against this theory for most debris banding is that thrusting would not lead to such a systematic downstream pickup of material. Also, the banding is usually parallel to the crystal orientation of the ice and a thrust would tend to cut across any preferred crystal orientation. Boulton favours refreezing in the lee of small basal obstacles to explain sediment-layering at the base of cold glaciers. Pressure-melting and refreezing around a small protuberance would cause a layer of clear ice to form in the lee, jacking up any basal sediment which might be present. In a cold-base glacier, the layer sequence of clear ice and sediment would be retained, and more layers added at other protuberances downstream, until a considerable thickness of sediment-laden ice developed (Fig. 6.25). In temperate glaciers, pressure-melting would not allow preservation of the ice layers downstream, and therefore one would not expect a significant amount of subglacial debris to accumulate. Supraglacial transport therefore assumes greater importance in temperate glaciers.

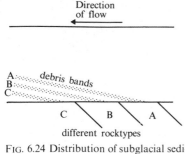

FIG. 6.24 Distribution of subglacial sediment in a cold-base glacier.

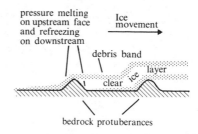

FIG. 6.25. Incorporation of subglacial debris by refreezing in the lee of obstructions.

Whether debris is deposited subglacially or supraglacially from a glacier has a significant influence on the characteristics of the resulting sediment. Subglacial deposition takes place from masses of basal ice which become stagnant, perhaps in the lee of very large obstacles, and which are separated from the moving ice above by a shear plane. Slow melting of the stagnant mass releases incorporated sediment layers, and water is expelled by the pressure of the overlying ice. The resulting deposit, so-called *lodgement till*, is extremely compacted due to the pressure under which it was laid down. Deposition of *melt-out till* takes place from stagnant ice-masses isolated by melting back of the glacier during retreat. Large masses of ice are often preserved beneath glacial or fluvial sediment, and take a considerable time to melt. Melt-out till is also deposited under pressure, though not as much as lodgement till, and is deposited so slowly that the internal, banded structure mentioned above is often preserved.

The most complex zone of deposition in a valley glacier system occurs at the terminus itself, where supraglacial and sub-glacial processes combine with fluvial processes in front of the glacier, to produce an extremely complex suite of sediments. Towards the snout, debris is brought to the ice surface along thrust planes and by ice melting to reveal former englacial debris to form *ablation till*. (Ablation is the term applied to melting of glacier ice.) Accumulating debris on the glacier protects the underlying ice from further melting, and where debris is brought up along thrusts this may lead to the development of ice-cored ridges (Fig. 6.26). Ablation till often becomes unstable because of its high water-content and unconsolidated nature, and flows from the glacier surface to accumulate in front of it as *flow till*, which is of course entirely unstratified.

From the above discussion of glacial transport and deposition, it is clear that glaciers do not transport material in a way similar to fluids. Debris is not entrained from the bed according to the hydraulic principles discussed in 6.1 but is frozen to the glacier sole in the lee of obstructions. Furthermore, a large amount of the glacier load is derived supraglacially, and is simply rafted in position, or internally because of fresh ice accumulation. Glaciers also show little or no tendency to sort sediment by preferential entrainment

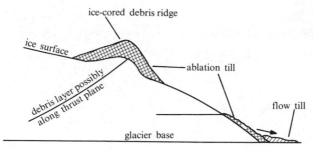

FIG. 6.26. Development of supraglacial tills at the glacier snout.

of smaller particles. Supraglacial sediment is rafted with little change except by sub-aerial weathering and, while subglacial sediment is crushed and comminuted in transport, material of all sizes is incorporated into the ice. The non-sorting, crushing nature of the glacier-sole environment is alluded to in section 7.4. Finally, glacial deposition is related less to the mechanics of glacier movement than to the simple melting-away of the transporting medium.

Further Reading

BOULTON, G. S. (1972a). 'Modern Arctic Glaciers as Depositional Models for former Ice Sheets', *Quarterly Journal of the Geological Society*, **128**, 361–93.

—— (1972b). 'Role of Thermal Regime in Glacial Sedimentation', *Institute of British Geographers Special Publication*, **4**, 1–19.

BUNTING, B. T. (1961). 'The Role of Seepage Moisture in Soil Formation, Slope Development and Stream Initiation', *American Journal of Science*, **259**, 503–18.

CARSON, M. A. (1971a). Op. cit. page 56.

CHEPIL, W. S. and WOODRUFF, N. P. (1963). 'The Physics of Wind Erosion and its Control', *Advances in Agronomy*, 211–300.

EMMETT, W. W. (1970). 'The Hydraulics of Overland Flow', *United States Geological Survey Professional Paper*, 662A, 68 pp.

HORTON, R. E. (1945). 'Erosional Development of Streams and their Drainage Basins: A Hydrophysical Approach to Quantitative Morphology', *Bulletin of the Geological Society of America*, **56**, 275–370.

KIRKBY, A. V. T. and KIRKBY, M. J. (1974). 'Surface Wash at the Semi-Arid Break of Slope', *Zeitschrift für Geomorphologie*, Supplementband **21**, 151–76.

LEOPOLD, L. D., WOLMAN, M. G. and MILLER, J. P. (1964). *Fluvial Processes in Geomorphology*, Freeman, New York.

MOSELEY, M. P. (1973). 'Rainsplash and the Convexity of Badland Divides', *Zeitschrift für Geomorphologie*, Supplementband **18**, 10–25.

—— (1974). 'Experimental Study of Rill Erosion', *Transactions of the American Society of Agricultural Engineers, Soil and Water Division*, 909–13.

PATERSON, W. S. B. (1969). *The Physics of Glaciers*, Oxford.

SCHUMM, S. A. (1956). 'Evolution of Drainage Systems and Slopes on Badlands at Perth Amboy, New Jersey', *Bulletin of the Geological Society of America*, **67**, 597–646.

WILSON, I. G. (1971). 'Desert Sandflow Basin as a Model for the Development of Ergs', *Geographical Journal*, **137**, 180–99.

7 Process Regimes and Time Aspects of Material Transport

7.1. Process regimes and a general sediment system

Throughout this book the mechanics and chemistry of sediment- and solute-transfer processes have been considered one at a time, with little attempt to unify them. In this chapter the spheres of influence of different groups of process will be examined and an attempt made to fit them together into a general system for material movement. This involves the interaction of processes and the way in which material is transported from one process to another.

Thus far, processes have been differentiated into slow and rapid mass movements and into fluid transfers. It is possible to define the system geographically rather than by process mechanics, and to delimit environments or *regimes* which are dominated by a specific group of processes. In some cases regimes are simply spatially defined, in others time scale must also be considered in order to define the boundary of a regime.

Briefly, there are three land-based regimes which are readily identifiable. First, there is a weathering regime which encompasses all processes of mechanical and chemical breakdown of rocks. These processes are mentioned only in briefest outline in chapter 2 but are the subject of many texts, for example Ollier (1969). Although weathering takes place all over the earth and within the regolith, it is dominant in specific parts of the system, particularly at the so-called weathering front, which is the more or less diffuse surface dividing intact rock below from weathering products above. There are two important cases, one in which removal of released weathered material takes place immediately and is therefore input-limited, and one in which material is not moved immediately but is released into a regolith cover lying on the bedrock. In other words, transport processes do not operate sufficiently quickly to remove weathered material. In the former case a bare rock-surface remains exposed to the atmosphere and in the latter there is a variable thickness of unconsolidated debris over bedrock. The influence of the weathering regime is extremely variable. On some unjointed bare rock surfaces it may only be a thin skin of chemically altered rock a millimetre or so in thickness, whereas in well-jointed rocks it may extend several metres from a bare rock-face. Alternatively it is a zone of variable thickness within the ground which separates fresh rock from the soil. Beneath a regolith the zone of active weathering may be quite thin and may quite justifiably be referred to as a weathering front. The change from completely fresh to completely weathered rock may take place in a matter of a few tens of centimetres or less. This is not to deny the continued operation of weathering within the regolith, but the changes are slow and small in comparison to those occurring at the weathering front itself (Fig. 7.1). Clearly this is a simplistic view of regoliths which is often not obtained in practice; it is used here only as a basis for discussion of process regimes. The weathering regime therefore exists as a thin skin over the earth with boundaries roughly parallel to the earth's surface. Almost all sediment and solutes must pass through this zone before entering a transport system.

A second regime of processes may be termed the *slopes* regime, in which material is moved in rapid and slow mass movements down the gravity

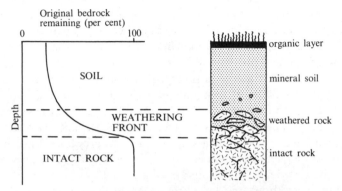

FIG. 7.1. Degree of weathering with depth in a soil.

gradient, and in slope wash processes. All the regolith above the weathering front is potentially part of the slopes regime, and the major driving force is the downslope component of the regolith's own weight. As suggested in Fig. 7.1, the distinction between the weathering and slopes regimes is much more clear cut than might be imagined, though continued weathering within a regolith might have a significant influence on mass-movement processes. The division between mass-movement processes and water-transfer processes on slopes is suppressed here, with a view to grouping processes together by area of influence. It must be borne in mind, however, that these processes are very different from the point of view of how material is entrained and the velocity at which it travels.

Finally, one can envisage a series of *fluid*-transfer regimes: water, air and ice. In these the driving force is primarily fluid stresses exerted at the bed of the flow. Water is of course the most important of these regimes, and the discussion will largely be confined to this. The boundaries between the slopes regime and the fluvial regime in a drainage basin are the stream bank and the channel-head regions. All material entering the river system must cross these boundaries, though it does so in a number of different ways. Slow-moving, diffuse, intergranular water, or throughflow, mainly carries solutes (chapter 3 and section 6.3). Overland flow, which is relatively rapid, is able to entrain sediment owing to stress exerted at the base of the flow,

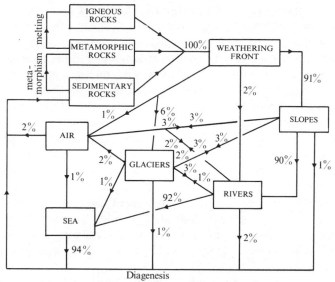

(a) Flowpaths in a general transport system

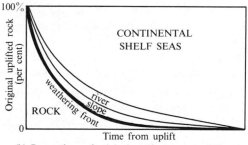

(b) Proportions of original uplifted rock in different
    process regimes

Fig. 7.2. A general transport system.

and to supply it directly as suspended load or bedload to the channel. Finally, the stream bank delivers sediment by processes such as undercutting and collapse. The relative importance of these ways in which material is supplied in terms of sediment yield depends on the hydrology of the drainage basin, which has already been alluded to in section 6.4.

These parts go to make up a complete sediment transfer system, such as a drainage basin, which acts as an organized entity. The regime boundaries are more or less distinct geographical surfaces with differing characteristics across which material is transmitted. There tends to be an abrupt change of transport mode on either side of a boundary, and there may be associated changes in material quality. Although these boundaries have been proposed as very clear divisions, they are to some extent gradational. As in all natural systems, sharp divisions should be regarded with suspicion.

The process regimes provide a convenient framework for a general system of material transport from the land surface (Fig. 7.2a). Its major features are that sediment passes through a weathering regime, a slopes regime and some fluid regime during transport to the sea. Solutes pass directly into a water transport system and tend to stay there. Of course, material flowpaths are never quite as simple as this. There is always the possibility of short-circuiting a regime or back-tracking in the system to a previous one. Figure 7.2a shows some linkages which may occur during transport, but by no means all the possible flow routes are shown. Relative proportions of the material derived from weathering following specific routes are indicated in the diagram, though these are obviously intended as a rough guide only. After uplift, a land surface becomes exposed to weathering and transport processes. The proportion of the original uplifted rock in each regime through time is indicated by Fig. 7.2b, which shows the general trend of original rock through the weathering and transport regimes until it is almost entirely returned to a marine depositional environment.

In areas where permanent stream-networks exist, the drainage basin provides a convenient natural unit within which to study transport processes. The drainage basin is a well-defined area draining to a single outlet, through which effectively all material derived in the area by weathering must pass (neglecting aeolian transfer, which is usually small in areas of permanent drainage). The system consists of a weathering front with overlying regolith and slopes which drain towards a permanent channel system (Fig. 7.3). An important aspect of drainage basins is the tendency for increased concentration and organization of drainage and material transfer as the outlet is approached. For example, overland flow and subsurface flow concentrate into rills and pipes respectively, and these concentrate further into a channel system. The implications of concentration in terms of how quickly material moves through parts of the system is discussed in section 7.2.

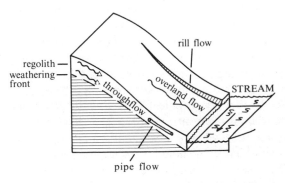

FIG. 7.3. The drainage basin system.

Growth and interaction of regimes

Although it was suggested above that process regimes are fairly discrete, some interactions must take place between them, especially during growth on a new surface. In a newly developing system, such as a drainage basin initiated on a new landscape, process regimes must be changing their sphere of influence until they occupy an equilibrium position with respect to one another. This involves modification of regime boundaries through time, and some of these must inevitably involve the growth of one regime at the expense of others. When and if equilibrium is established, fluctuations may occur about mean positions but these will, on the whole, cancel each other out.

The growth of a permanent channel network of rills or streams provides good illustration of an approach to equilibrium on a newly exposed surface. On exposure, the surface is under the influence of weathering. But water in the system becomes more dominant through time and eventually leads to the development of a channel network. As mentioned in section 6.4, the type of channel developed depends on hydrology. Where overland flow is dominant, in low infiltration-capacity soils, a permanent rill system with ephemeral flow will occur. In low infiltration-capacity or high surface-erodibility areas, channel initiation is possible anywhere outside the zone of no sheet erosion, adjacent to all divides. Thus channelling continues to increase until the divide areas are reduced to about two 'no erosion' zones in width. Since this zone is small, a matter of a few centimetres to less than a metre in naturally developing systems (Schumm, 1956; Segerstrom 1950) and in laboratory experiments (Flint, 1973; Moseley, 1974), extremely dense networks result. Where infiltration capacity is rarely exceeded by rainfall, overland flow is restricted to places where the soil becomes completely saturated; for example, percolines, topographic hollows and breaks in slope (section 6.4). Overland flow is therefore possible only in limited zones and, assuming it is a prerequisite for channel cutting, channel density will be much lower. This is the situation commonly found in humid areas. Growth of channels will continue until all areas which capture enough subsurface flow to generate overland flow have become cut. Permeable bedrock, such as the Chalk in England, shows the extreme for subsurface flow capacity, being almost entirely devoid of active surface-flow lines. Few data exist upon the rate at which drainage networks develop, but what are available suggest that growth is relatively rapid. Drainage density tends to approach its limit or *equilibrium drainage density* quickly and then to remain constant (Fig. 7.4). This is not to say that the network shows no further change, but that a newly initiated channel is roughly balanced by infilling of other channel reaches. (It should be noted here that a channel is taken to indicate an active channel which regularly takes flow, for the purposes of equilibrium drainage density.) The amount of surface runoff is paramount in determining the rate and extent of channel development, as well as other closely linked factors such as soil erodibility and vegetation. Network growth rate is illustrated here by just three examples. Laboratory simulations, where surface runoff is very great, reach equilibrium in a few hours (Fig. 7.4a). Ephemeral rill systems on newly exposed ground, on the other hand, take months or years to attain equilibrium drainage density (Schumm, 1956). Finally, an interesting study by Ruhe (1952) has shown that humid stream-systems develop much more slowly, though still very quickly in geomorphic terms. Ruhe examined the characteristics of the drainage net on glacial till sheets of different ages in Iowa, and his data show a rapid increase in drainage density in the first 20 000 years after exposure, with little subsequent growth (Fig. 7.4b).

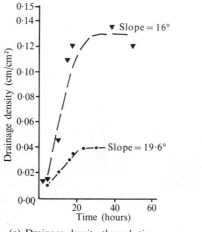

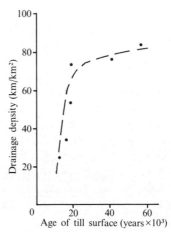

(a) Drainage density through time in laboratory experiments (plotted from data in Flint, 1973)

(b) Drainage density through time on glacial tills (after Leopold, Wolman and Miller, 1964, from data in Ruhe, 1952)

FIG. 7.4. Development of drainage networks through time.

The development of a regolith over a new bedrock surface also exemplifies interaction between regimes in attainment of equilibrium. The rate of regolith accumulation is a result of the difference between weathering and removal by mass movement and slopewash transport processes. But also, as a regolith develops, the rate of weathering at the weathering front declines, owing to the insulation effect of the accumulated debris. Thus the thickness of the regolith or slopes regime depends on interactions at two surfaces, the ground and the weathering front. As the slope regime increases, the rate of weathering declines until the regolith achieves an equilibrium depth at which input by weathering is balanced by all the losses. The equilibrium depth is zero when transport from the surface is potentially faster than the rate of supply of weathered material. In this case removal is said to be *weathering-limited*. In the alternative situation, where a regolith accumulates, removal is said to be *transport-limited*.

Explicit in the above examples is the concept of an equilibrium between regimes, which involves a dynamic interchange of material and space about a mean position. The form of interaction between parts of the whole transport system depends upon process capacity, weathering susceptibility of bedrock, and transport susceptibility of the weathering-derived material. To reiterate, interactions tend either to be input-limited, that is the capacity of transport processes exceeds the rate at which sediment is made available by weathering, or they tend to be process-limited when transporting capacity was initially less than the rate of release of material by weathering. If the boundary between weathering processes and slope transport processes is weathering-limited, then a bare rock-slope results. Since most transport processes become more effective with increasing slope angle, it follows that steeper slopes are more likely to be devoid of any sediment cover. Also, hard bedrocks will weather more slowly and are more likely to be weathering-limited. It is certainly true that there is a strong tendency for steeper and harder rock slopes to be soil-free. On the other hand, a transport-limited boundary will result in soil accumulation, because weathering releases material faster than it can be carried away. If equilibrium is to be approached, the rate of weathering must be increasingly inhibited by the accumulation of more soil, until input by weathering to the soil layer is just balanced by transport. These arguments may be extended to the boundary between slopes and rivers where, if transport capacity is greater than the rate of supply of sediment, the boundary is erosion-limited and a bedrock channel results. Alternatively, flood-plain accumulation is a reflection of sediment supply being in excess of transporting capacity.

7.2. Time aspects of material in transport

The concept of *residence time* in parts of the hydrologic cycle is well known. Residence time is the mean length of time a water molecule stays in a particular part of that system: rivers, groundwater, the atmosphere or the oceans. It is calculated very simply by dividing the volume of water in a part of the system by the rate at which water passes through it. Residence time is of course lengthened by increasing the volume of part of a system or by decreasing the rate at which water flows through it.

Residence time can also be applied, with reservations, to material in transport systems, and a few examples will be examined here to show the relative lengths of time material is held in parts of a transport system. The calculations below are unavoidably hypothetical because very few measurements concerning residence time have ever been made.

Example 1: The regolith

Assuming equilibrium regolith-thickness at about 500 mm (not unrealistic over many upland slopes in Britain, for example) and assuming that the average world condition of about 30 per cent solution and 70 per cent sediment transfer takes place from the regolith, calculations for residence time in the regolith are as shown in Table 7.1. They are based on average rates of denudation in different regions and upon rates of loss from a regolith by specific processes. The rates used are largely average values calculated from tables in Young (1974). The main conclusion is that residence time in the regolith is of the order of many thousands of years. Also, rates calculated by over-all denudation-rates and by specific process-rates are comparable for temperate regions, especially when it is recognized that many slopes will be influenced simultaneously by soil creep and landslides (Table 7.1). It must be borne in mind that the complicating effect of solutional loss has been largely neglected because of the assumption that solute loss is instantaneous once released by weathering and is entirely from the base of the regolith.

Example 2: A hypothetical river network

Residence time within a river network is considered for a nominal drainage area of 100 km^2 with a main-stem channel length of 16 km and total channel length of 300 km. This is based on a relationship between drainage area and channel length for the Shenandoah Valley in the U.S.A., which also seems to hold for larger world rivers (Leopold, Wolman and Miller, 1964). An average channel width of six metres, with one metre depth of sediment within it and no flood plain, is assumed. The hypothetical nature of this calculation need not be emphasized further! The calculations are shown in Table 7.2 and the residence time derived for this situation is 48 years in the main-stem channel. Even allowing for the speculative nature of the calculations it must be appreciated that residence time in the regolith and in a river system are on completely different scales. This is a reflection of the concentration of transfer within a drainage basin with change-over to the fluvial system from the regolith. Again it should be noted that solutes have been neglected, the assumption being that their residence time is comparable to that of the water in the system, which is very short.

If a floodplain exists in a river system then the volume of sediment which can be regarded as in transfer is increased, with a concomitant increase in residence time. However, the total volume of sediment in a river system including a floodplain must be considerably less than the volume in the regolith within the basin, simply because the latter occupies so much greater an area. Hence, the existence of a floodplain will probably not affect the magnitude of the difference between regolith and river-system residence time very much. A rough calculation is included in Table 7.2 for the floodplain situation, assuming that all material passes through it. This is clearly an upper limit to residence time, since in an equilibrium situation where the floodplain is of stable volume, only a small percentage of all material will pass through it. The justification for this is that most sediment transported in rivers remains in suspension, and overbank flooding, which would give an opportunity for this sediment to be deposited on the floodplain, is a rare occurrence. Also Leopold, Wolman and Miller (1964) calculated that a par-

Table 7.1
Residence time for material in a hypothetical regolith

Regolith thickness—500 mm (in equilibrium).
Weathering front thickness—200 mm (in equilibrium).
Sediment: Solute ratio in rivers draining from area—70 : 30.

Residence time using approximate European average denudation rate of 20 mm/1000 years (a depth of material stripped from the land surface in unit time).

   Rate of denudation by solute loss—6 mm/1000 years.
   Rate of denudation by sediment loss—14 mm/1000 years.

$$\text{Residence time for sediment in the regolith} = \frac{\text{volume of system}}{\text{rate of transfer through system}}$$

$$= \frac{50 \times 1000}{14} \text{ yrs} = 36\,000 \text{ years.}$$

$$\text{Residence time for sediment and solutes in the weathering front} = \frac{\text{volume of system}}{\text{rate of transfer through system}}$$

$$= \frac{200 \times 1000}{20} = 10\,000 \text{ years.}$$

(The calculation makes the gross assumption that all solutes are derived and lost directly from the weathering front and that all sediment passes through the regolith.)
Residence time using world average, high-relief, denudation rate of 440 mm/1000 yrs:

$$\frac{500}{400} \times 1000 = 1600 \text{ years for sediment.}$$

Residence time using world average denudation rate of 52 mm/1000 yrs:

$$\frac{500}{42} \times 1000 = 14\,000 \text{ years for sediment.}$$

Residence time using tropical soil depth and denudation rate on low slopes:

$$\frac{10\,000 \times 1000}{1} = 10 \text{ million years.}$$

Residence time calculated from loss by specific processes
Soil creep (temperate average)—4 cm^3/cm/yr, which yields a rate of denudation on a 50-metre-long slope with 500 mm thickness of regolith of 8 mm/1000 yrs.

Residence time for sediment in the regolith:

$$\frac{500 \times 1000}{8} \text{ yrs} = 60\,000 \text{ years for a sediment particle starting from the top of the slope, or an average of } 30\,000 \text{ years.}$$

Landslides (temperate estimate)—11 mm/1000 yrs.
Residence time for sediment in the regolith:

$$\frac{500}{11} \times 1000 = 43\,000 \text{ years.}$$

Table 7.2

Residence time for materials in a hypothetical river

Drainage area—100 km²; total channel length = 300 km, main stem channel length = 16 km.

Channel width—6 m (average) } for main stem channel
Sediment depth in channel—1 m }

volume of sediment in the main stem channel system = (16 × 1000) × 6 × 1
$$= 96\,000 \text{ m}^3$$

Rate of denudation by sediment loss assumed to be 30 tonnes/km²/yr (from a figure for the Seine in Holeman (1968)). Assuming a dry unit weight of 1·5 tonnes/m³, the volume of sediment is 20 m³/km/yr (20 mm/1000 yrs). The yearly loss of sediment from the drainage basin is therefore:

(drainage basin area × yearly loss per unit area)
$$= (100 × 20) \text{ m}^3/\text{yr} = 2000 \text{ m}^3/\text{yr}.$$

Residence time $= \dfrac{96\,000}{2000} = 48$ years in the main stem channel.

For the total network length, making the same assumptions,

Residence time $= \dfrac{1\,800\,000}{2000} = 900$ years.

including a flood plain 100 m wide × 5 m deep

Residence time $= \dfrac{16\,000 × 5000}{2000} = 4000$ years for the main stem channel.

---

ticle of sediment deposited on the inside of a meander bend in Seneca Creek (Maryland) would have to remain in the floodplain 11 000 years before being eroded by the passage of the next meander. This would give extremely long residence times for sediment which was repeatedly caught on each meander bend, and so it seems probable that only a fraction of the total load is actually deposited in the floodplain. It may be more meaningful to view residence time in the river system as a distribution (Fig. 7.5) rather than as an average time.

As a generalization, it can be said that the regolith is the buffering agent in the drainage basin system. In other words, it is transfer from the regolith

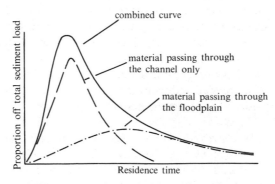

FIG. 7.5. Distribution of residence time for sediment in river systems.

which ultimately controls transfer through the system. The justification for this statement is twofold. First, most of the land surface is transport-limited; bare rock-surfaces are rare. Hence, regolith accumulation generally takes place to slow down the rate of bedrock weathering to a rate at which it can be moved from the slope. Secondly, most of the time, rivers carry material at less than full capacity, and so the rate of river transport is governed by supply of material as well as by the mechanics of fluid transport.

7.3. Magnitude and frequency of transfer processes

Two further aspects of transport processes need to be considered in assessing their importance, which are: how often they occur, and how much material is moved by them when they do? The former is relevant only to processes having a threshold level of applied stress, below which they do not occur, and the latter is dependent on the magnitude of the force applied to the material being transported. Processes can thus be split into those continuously operating, such as soil creep and solution, and threshold-controlled processes, such as landslides and sediment movement by fluids. Continuously operating processes naturally cause constant transport, though rate increases with increasing applied stress. Threshold processes, however, only increase in importance with applied stresses above the critical level.

Magnitude and frequency of transport processes are evidently related to magnitude and frequency of controlling events. In the short term the most significant events are undoubtedly meteorologic and climatic. Usually rainfall is the key meteorological factor, since water influences many mass-movement processes and of course water-transport processes. Other factors are significant, however; for example, temperature has some bearing on many creep processes and of course wind speed controls aeolian transport. The question arises as to what size of recurrent meteorological event is of most significance in terms of total sediment-transport by a particular process. So-called catastrophic floods (clearly linked to rare rainfall events) cause massive rates of transport, but do they occur frequently enough to be of measurable importance in terms of total transport? Another issue, not discussed here at all because it is beyond the scope of this book, is the significance of events in determining the *form* of the transport system. It should be remembered that events which carry the most material overall may not be those to which form is adjusted.

In the limited amount of space available here we shall look at magnitude and frequency of suspended-sediment loss from river systems, since this reflects the magnitude of all sediment-derivation processes in the river basin. The frequency of floods on river systems, intimately linked to rainfall, tends to be log-normally distributed with a long tail towards the less frequent events. This is illustrated by Fig. 7.6, which shows the relationship between flood size, expressed as a ratio of the mean annual flood, and recurrence for some Pennsylvanian rivers (mean annual flood is the mean discharge of the

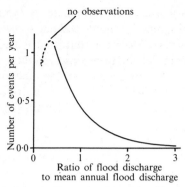

FIG. 7.6. Frequency of floods on some Pennsylvanian rivers. (Replotted from data in Leopold, Wolman and Miller, 1964.)

yearly maximum floods over a number of years). The distribution is strongly skewed towards more frequent events. The mean annual flood recurs about once in 2·33 years, while the average yearly flood is about 0·3–0·4 times the mean annual flood. The recurrence interval of large floods declines exponentially with flood size. The size of most important events is determined by the product of this relationship and the relationship between transport rate and flood discharge (Fig. 7.7).

Wolman and Miller (1960) calculated for rivers with a wide variety of discharge and sediment load characteristics that almost without exception 98 to 99 per cent of the total suspended load is carried in storms recurring more frequently than every ten years. Most in fact carry 80 to 95 per cent of their suspended load in events recurring more frequently than once a year. The so-called catastrophic events, which should perhaps be regarded as systematic but occurring in the tail of the frequency distribution, are of very minor over-all importance. On the other hand the smaller events occurring on average 320 days per year (87 per cent of the time) on the Colorado River transported only 40 per cent of the total load. The other 60 per cent was transported on only 45 days. Thus it is the moderate-sized events, with

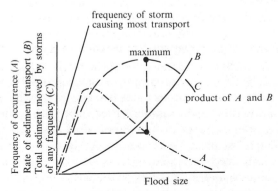

FIG. 7.7. Determination of the most significant flood size in terms of total sediment transport.

recurrence intervals of a year or less, which are most important in suspended sediment movement. Figure 7.8 shows the percentage of total annual load against the number of days per year that load is likely to be exceeded for two rivers. Both show similar relationships even though their sediment loads are very different.

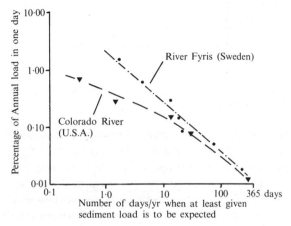

FIG. 7.8. Expected minimum daily sediment load distribution on two rivers. (Calculated from data in Douglas, 1964.)

It should be mentioned that meteorological factors are not necessarily the only ones controlling the input of stress to a transfer system. Longer-term factors can be important, depending on time-scale. For example, tectonic activity has already been shown to be a significant landslide control in some places (section 4.7 and Fig. 4.12). The frequency graph for earthquakes needs to be superimposed upon that of rainfall to ascertain the relationship between frequency and total transport by landslides in a tectonically unstable area. Slow changes in material strength through time by weathering will influence their response to applied stress, and also need to be considered over long time periods. Finally, the effect of climatic change on a long time-scale will be to modify the frequency curve for meteorological events.

7.4. Changing grainsize characteristics throughout transport

This section looks briefly at changing sediment mechanical properties during transport and concentrates on grainsize distribution. The initial grainsize distribution of sediment derived by weathering from fresh rock is a function of rock properties and weathering process. The most important weathering process factor is the distinction between chemical and mechanical action. Complete chemical degradation of silicate rocks, for example, will yield clay minerals and predominantly clay-sized material. Dominant mechanical breakdown yields an initial grainsize distribution closely related to joint or discontinuity spacing in the rock, but subsequent development

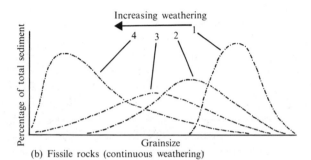

(a) Granular rocks (discontinuous weathering)

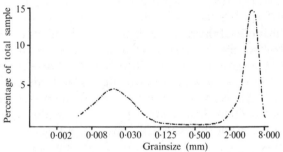

(b) Fissile rocks (continuous weathering)

(c) Breakdown of quartz particles after 16 hours grinding in a ball mill (after Gaudin, 1926, in Dreimanis and Vargners, 1972)

FIG. 7.9. Grainsize distribution trends through weathering.

will depend on how the rock responds to the weathering process. Response may be simplified by considering two cases: perfectly granular rocks and perfectly fissile rocks. A perfectly granular rock, for example a sandstone, will tend to break down into small constituent particles but will be resistant to splitting of the larger clasts. Weathering is therefore a gradual reduction in clast size to produce fine-grained debris. In short, they *wear*. The trend in size distribution, shown in Fig. 7.9a, is initially towards bimodality and finally to a unimodal, fine-grained sediment. Thus there is a *discontinuous* distribution of grainsize (discontinuous weathering) within the weathered material. Fissile rocks such as shale, on the other hand, are just as likely to split into intermediate-sized clasts as they are to disintegrate into fine grains. The process could be called *breakage*. Consequently there is a more *continuous* distribution of grainsize throughout weathering (continuous weathering, Fig. 7.9b). Little direct evidence for these trends in weathering exists, but experiments on rock breakage in ball mills illustrate the principle. Quartz particles developed a bimodal size distribution in one such experiment (Fig. 7.9c) and other hard siliceous rocks have shown similar behaviour. These materials, while not strictly granular, tend to wear rather than break. Hard shales, in contrast, did not tend to develop bimodality when treated in a similar manner (Dreimanis and Vargners, 1972).

When the influence of transport processes on grainsize change is considered, one has to bear in mind not only progressive grainsize reduction by wear or breakage but also changes due to selectivity of the transport process. Wear and breakage take place in all transporting environments by grain impacts or by crushing under load, and grainsize reduction can be regarded as a universal trend. Selective transport and deposition (*sorting*) are, however, related to the mechanics of the transport process. Generally speaking, mass movements do not sort material, because grains move collectively and not as individuals. This is also generally true of glacial transport, where once again particles are moved with roughly equal status, independent of size.

Transporting fluids, in contrast, possess the ability to sort. This is because there is a threshold bed stress for each grainsize at which it begins to move, and only particles smaller than the critical size will be moved by the flow. Sorting is thus a result of some material being left behind and of different grainsizes moving at different velocities. The big break in velocity is probably related to the transition between bedload movement, which is slow, and suspended load movement (or saltated load in the case of wind transport), which is faster. This distinction is not, however, always very sharp. In addition to sorting at inception of transport there is also sorting during waning flow, since coarser material will be dropped first as bed stress declines.

Sediments in non-sorting environments tend, not surprisingly, to be poorly sorted. In mass movements the material reflects closely the size distribution of the weathered material, with some limited crushing and breaking along the shear planes of landslides. Glacially transported sediments are also poorly sorted, often containing material ranging from clay-size grains to boulders in the same deposit. The glacial environment is a severe crushing environment, however, causing decreasing grainsize away from the source area. Figure 7.10a shows grainsize curves for two glacial sediments from near Dublin, Eire, one which has been transported hundreds of kilometres from the north, the other which has only come about ten kilometres from the Wicklow Mountains. Both curves have similar form, with a wide size-range and approximately straight-line relationship, but the material which has been transported further is much finer in size. It appears that transport simply displaces the curve to the left.

Sediment in fluid transport tends to be well sorted. This applies equally to the material in transport and to the material settling from flow at any point. Grainsize curves for desert-dune sand and river-bed sediment are shown in Fig. 7.10b. Both are comparatively well sorted, though the river silt is much less so. This is attributable to a number of reasons. First, the range of sizes available for transport by desert winds is generally less than for rivers (recall that most desert sands are derived from river deposits which have already been comminuted and sorted), and the range *transportable* by wind is any-way less. Secondly, at-a-point sorting in a river channel is complicated by continued supply of new and unsorted sediment from the banks and from tributaries.

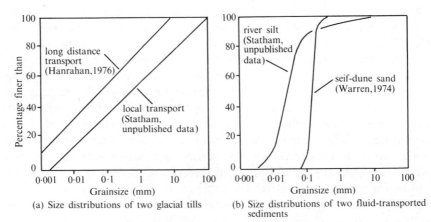

(a) Size distributions of two glacial tills

(b) Size distributions of two fluid-transported sediments

FIG. 7.10. Grainsize distribution curves for selected environments.

Take a river system as an example of how fluid transport changes grain-size. Wear, breakage and sorting all act to cause change downstream. In an equilibrium channel, where all input sediment is output from the reach, the sorting effect alone cannot cause an over-all reduction in grainsize down-stream. This is fairly apparent because all material input to the channel must eventually pass out of the channel mouth and so ultimately the size of the transported material will be the same as the derived. Over-all grainsize reduction is in fact due to wear and breakage, and this has already been shown for a glacial outwash stream, in Fig. 6.13. Grainsize declined down-stream and the percentage of harder rocktype increased, showing the breakage had reduced the occurrence of less resistant ones. Over longer distances, the Mississippi shows sediment changes very well (Fig. 7.11). Grainsize declines logarithmically with the logarithm of distance, which means that the rate of decline decreases downstream. Sorting also increases rapidly in the first few kilometres and then much more slowly afterwards (Leopold, Wolman and Miller, 1964). The trends are progressive over the whole length of the river but are complicated by tributaries entering the main channel, where grainsize tends to increase temporarily.

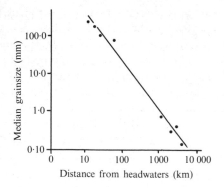

FIG. 7.11. Variation in median grainsize along the Mississippi. (From data in Leopold, Wolman and Miller, 1964.)

Only one sediment property has been singled out for attention here, namely grainsize. But all sediment properties—grain shape, roughness, surface form, mineralogy and so on—change through transport. The point which this section has been able only to introduce is that sediment is not an unchanging material. It undergoes constant alteration due to wear, breakage, selection and chemical alteration, and the product delivered to the mouth of a large river, for example, bears little resemblance to the material derived at weathering fronts in the far reaches of the basin. Its final state is a reflection of the whole range of external influences which have been brought to bear upon it in the sediment-transport system.

7.5. The relationship of material transport to other earth surface processes

This final section is an attempt to look at the weathering transport system in the context of other systems at the earth's surface with which it interacts. This was alluded to in chapter 1 (Fig. 1.1) but will be expanded somewhat here to demonstrate the importance of these interactions, in terms of earth surface processes as a whole.

The earth's surface is the theatre in which weathering and transport processes act. Only a very thin skin of rock indeed is influenced by these processes, but their action to a large degree determines the character of the landscape. At the most fundamental level, earth surfaces are either bedrock exposures or are soil covered; the latter being by far the most common situation. Bedrock exposure only occurs when the potential capacity for transport exceeds the rate of release of material by weathering. This may be due to very resistant bedrock or to high transport-capacities, the latter including gravitational processes on steep slopes or, possibly, rapid fluid flow. The fact that bedrock exposure forms a minute part of the earth's surface can be taken as an indication that a balance between input of material to a surface by weathering and output by transport usually requires a regolith to accumulate. Reasons for this are twofold, and the distinction between them a reflection of rocktype. First, a bedrock which breaks down mechanically to small particles may require a build-up of weathered material to insulate the surface from the direct effect of subaerial processes such as frost-action. In this way, the rate of weathering is reduced to establish an equilibrium between transport and weathering. Alternatively, sediment derived from a resistant rocktype, which breaks down on exposure into large, joint-determined particles, may be too coarse to be transported. Thus it must remain on the surface until it is weathered to a more easily transportable state before equilibrium is established. In both cases, an increase in regolith thickness is responsible for a reduction in direct mechanical attack on the bedrock and an increase in chemical weathering.

The importance of regolith or soil in the landscape cannot be over-emphasized. It is the final veneer which completes the land surface, and the layer in which many important weathering and transport processes operate. In the soil, one finds three systems juxtaposed and interacting with each other: the weathering-transport system, the plant-colonization and re-generation system, and the hydrological cycle. Probably the most important aspects of these interacting systems are the movement of water over the earth's surface and the derivation of plant nutrients. By considering how a bare surface may develop into one with complete soil and plant cover, it is possible to appreciate how interactions between the systems take place.

An initial bare rock-surface is open to attack by subaerial mechanical processes such as freeze–thaw and the effects of insolation, and to some chemical processes due to rainfall. This alone may be sufficient to cause surface breakdown and to build up a sediment layer. However, many more-resistant rocktypes, especially if they are poorly jointed, will be little affected by these processes and the all-important first stage of accumulation must be by other means. Some plants, for example lichens and mosses, are able to colonize bare rock-surfaces which are unweathered or which have only been very slightly altered by chemical weathering. Lichens especially are able to extract nutrients from rock surfaces by hydrogen ion substitution and therefore to aid chemical weathering. Subsequent accumulation of decaying plant material as humus, along with mineral matter from the bedrock, may be sufficient to form a layer in which other plants may colonize. Whatever form the first stage takes, it is most important because it aids water retention at the bedrock surface, thus enhancing chemical weathering and the capacity for plant growth. After the initial stage, the necessary mechanisms exist for full-scale plant colonization, namely available water for transpiration, and nutrient extraction. Development towards a complete vegetation cover involves three very important modifications to the processes operating over the bedrock surface. First, vegetation increases the amount of rainwater infiltrating into the soil and reduces overland flow, which serves to diminish soil loss by wash processes and to increase chemical weathering within and beneath the soil. Secondly, bedrock is increasingly insulated against the direct effects of subaerial processes, which reduces mechanical breakdown and may reduce the over-all rate of weathering if these processes had a severe effect on the original exposed rock surface. Thirdly, accumulating decayed plant material leads to the development of a humus layer. This is a very important aspect of soils indeed, since humus decay produces complex organic acids, important in chemical weathering and nutrient derivation (chapter 3), and stimulates soil fauna whose activities control carbon dioxide levels in the soil and, in turn, pH. Decaying plants also return nutrients absorbed in them to the soil.

The interaction between the plant-regeneration system and the weathering-transport system is most obviously demonstrated by chemical processes. Chemical weathering, aided by organically derived reagents, is the source of new plant nutrients and solutional transport load. Chemical weathering processes also release new material from bedrock to replace that lost by transport. Less obvious, but still important, is the relationship between chemical weathering and mass-movement processes. As soil strength declines through weathering, the probability of mass movement increases, with a resulting increase in soil loss. Slow mass-movements cause a continuous discharge of soil from a slope and, if a constant soil depth is to be maintained, the release of sediment from bedrock must be sufficient to replace that lost by solution and slow mass-movement. A similar situation would apply to removal at the surface by wash processes. Landslides and other rapid mass-movements are related rather differently. If weathering of the soil reduces strength sufficiently for it to become unstable, then landsliding will remove it to expose a new bedrock surface. The processes of plant colonization will then start anew, and soil development can be considered to be cyclic.

To summarize, the soil is a zone of very complex interaction. The most important part of this system is, however, the increase of infiltration accompanying vegetation colonization combined with all the processes of vegetation decay and plant regeneration, which controls soil chemistry, nutrient derivation and solutional loss. Figure 7.12 is intended to show how closely hydrology (and hence, indirectly, climate), vegetation and nutrient derivation, weathering and transport are linked within the soil, though clearly by no means all possible linkages are shown. The diagram serves to demonstrate the implications of change in any factor for the whole system. For example, a change towards a drier climate will have a detrimental effect on vegetation growth and will work through the system as a change in soil chemistry and an increase in the proportion of rainfall as overland flow. The latter factor will be responsible for increased slopewash and hence soil erosion. What applies to natural changes in the system will also apply to those initiated by man. Agricultural practices which leave unvegetated soil surfaces exposed for prolonged periods run the risk of increased soil erosion by slopewash and wind owing to increased overland flow and diminished surface protection. Indeed, devegetation has resulted in catastrophic soil loss in many parts of the world. Figure 7.12 also shows that failure to allow decaying plant material to return to the soil will modify soil chemistry and will

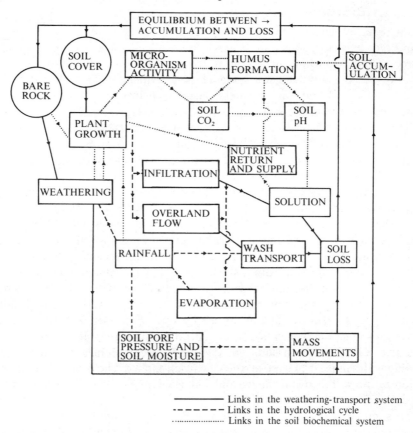

Links in the weathering-transport system
Links in the hydrological cycle
Links in the soil biochemical system

Fɪɢ. 7.12. Interactions between the plant nutrient cycle, hydrologic cycle and weathering-transport system within the soil.

lead ultimately to a reduction in the available nutrients and hence in soil fertility. Many other examples could be cited, but those given above are sufficient to show that no part of the soil system operates in isolation. In this light, material transport may be seen as one of an interrelated group of earth's surface processes which in the short term form a soil system developing towards equilibrium. Ultimately, however, these same processes are responsible for the steady erosion of the land surface. On this longer time-scale, equilibrium can never exist; steady, inexorable change by reduction of the land surface is inevitable.

Further Reading

Lᴇᴏᴘᴏʟᴅ, L. B., Wᴏʟᴍᴀɴ, M. G. and Mɪʟʟᴇʀ, J. P. (1964). Op. cit. page 156 above.
Wᴏʟᴍᴀɴ, M. G. and Mɪʟʟᴇʀ, J. P. (1960). 'Magnitude and Frequency of Forces in Geomorphic Processes', *Journal of Geology*, **68**, 54–74.

References

ALLEN, J. R. L. (1969). 'Maximum Slope Angle Attainable by Surfaces Underlain by Bulked Equal Spheroids with Variable Dimensional Ordering', *Bulletin of the Geological Society of America*, **80**, 1923–30.

—— (1970). 'The Angle of Initial Yield of Haphazard Assemblages of Equal Spheres in Bulk', *Geologie en Mijnbouw*, **49**, 13–21.

ATKINSON, R. J. C. (1957). 'Worms and Weathering', *Antiquity*, **31**, 219–33.

BAGNOLD, R. A. (1953). *The Physics of Blown Sand and Desert Dunes*, Methuen, London.

BERNER, R. A. (1971). *Principles of Sedimentary Geochemistry*, McGraw-Hill, New York.

BISHOP, A. W. (1966). 'The Strength of Soils as Engineering Materials', *Géotechnique*, **16**, 91–130.

—— (1973). Op. cit. page 100 above.

BJERRUM, L. and JØRSTAD, F. (1968). 'The Stability of Rock Slopes in Norway', *Norwegian Geotechnical Institute Publication*, **79**, 1–11.

BLIGHT, G. E. (1969). 'Foundation Failures of Four Rockfill Slopes', *Journal of the Soil Mechanics and Foundation Division Proceedings of the American Society of Civil Engineers*, **95**, 743–67.

BOULTON, G. S. (1967). 'The Development of a Complex Supraglacial Moraine at the Margin of Søobreen, Ny Friesland, Vestspitzbergen', *Journal of Glaciology*, **6**, 717–35.

—— (1970a). 'On the Origin and Transport of Englacial Debris in Svalbard, Glaciers', *Journal of Glaciology*, **9**, 231–45.

—— (1970b). 'On the Deposition of Subglacial Meltout Tills at the Margins of Certain Svalbard Glaciers', *Journal of Glaciology*, **9**, 231–5.

—— (1972a). Op. cit. page 156 above.

—— (1972b). Op. cit. page 156 above.

BRAMMER, H. (1971). 'Coatings in Seasonally Flooded Soils', *Geoderma*, **6**, 5–16.

BREWER, R. (1968). 'Clay Illuviation as a Factor in Particle Size Differentiation in Soil Profiles', *Transactions of the 9th International Congress of Soil Science, Adelaide*, 489–99.

BROWNING, J. M. (1973). 'A Catastrophic Rockslide at Mount Huascaran, North Central Peru', *Bulletin of the American Association of Petroleum Geologists*, **57**, 1335–41.

BUCKMAN, H. O. and BRADY, N. C. (1950). *The Nature and Properties of Soils*, Macmillan, New York.

BUNTING, B. T. (1961). Op. cit. page 156 above.

—— (1964). 'Slope Development and Soil Formation on Some British Sandstones', in 'Slope Profiles—a Symposium', *Geographical Journal*, **130**, 73–9.

CABRERA, J. G. and SMALLEY, I. J. (1973). Op. cit. page 100 above.

CAINE, T. N. (1963). 'Movement of Low Angled Scree Slopes in the Lake District, Northern England', *Revue Géomorphologie Dynamique*, 172–8.

CARBONNEL, J. P. and MEYBECK, M. (1975). 'Quality Variations of the Mekong River at Phnom Penh, Cambodia, and Chemical Transport in the Mekong Basin', *Journal of Hydrology*, **27**, 249–65.

CARRIGY, M. A. (1970). 'Experiments on the Angle of Repose of Granular Materials', *Sedimentology*, **14**, 140–58.

CARSON, M. A. (1971a). Op. cit. page 56 above.

—— (1971b). 'Application of the Concept of Threshold Slopes to the Laramie Mountains, Wyoming', *Institute of British Geographers Special Publication*, **3**, 31–49.

—— and PETLEY, D. J. (1970). Op. cit. page 100 above.

——, TAYLOR, C. H. and GREY, B. J. (1973). 'Sediment Production in a Small Appalachian Watershed during Spring Runoff: the Eaton Basin', *Canadian Journal of Earth Sciences*, **10**, 1707–34.

CHANDLER, R. J. (1969). 'The Effect of Weathering on the Shear Strength Properties of Keuper Marl', *Géotechnique*, **22**, 403–31.

—— (1970a). Op. cit. page 100 above.

—— (1970b). 'Solifluction on Low-Angled Slopes in Northamptonshire', *Quarterly Journal of Engineering Geology*, **3**, 65–9.

—— (1972a). Op. cit. page 100 above.

—— (1972b). 'Lias Clay: Weathering Processes and Their Effect on Shear Strength', *Géotechnique*, **22**, 403–31.

—— (1973). 'The Inclination of Talus, Arctic Talus Terraces and other Slopes Composed of Granular Materials', *Journal of Geology*, **81**, 1–14.

—— (1974). 'Lias Clay: Long Term Stability of Cutting Slopes', *Géotechnique*, **24**, 21–38.

CHEPIL, W. S. and WOODRUFF, N. P. (1963). Op. cit. page 156 above.

CROFTS, R. (1971). 'Sand Movement in the Emlybeg Dunes, Co. Mayo', *Journal of the Irish Naturalists*, **17**, 132–6.

DAVIS, S. N. (1964). 'Silica in Streams and Groundwater', *American Journal of Science*, **262**, 870–91.

DAVISON, C. (1888). Op. cit. page 113 above.

—— (1889). Op. cit. page 113 above.

DOUGLAS, I. (1964). 'Intensity and Periodicity in Denudation Processes with Special Reference to the Removal of Material in Solution by Rivers', *Zeitschrift für Geomorphologie*, **8**, 453–73.

DREW, D. P. (1972). 'Geomorphology of the Big Muddy Valley Area, Southern Saskatchewan, with Special Reference to the Occurrence of Piping', *22nd International Geophysical Congress, Montreal, Field Excursion Background Papers*, 197–211.

EINSTEIN, H. A. (1942). 'Formulae for the Transportation of Bedload', *Transactions of the American Society of Civil Engineers*, **107**, 561–97.

—— (1950). 'The Bedload Function for Sediment Transport in Open Channel Flow', *United States Department of Agriculture Technical Bulletin*, **1026.**

ELLISON, W. D. (1945). 'Some Effects of Raindrops and Surface Flow on Soil Erosion and Infiltration', *American Geophysical Union Transactions*, **26**, 415–29.

—— (1948). 'Studies of Raindrop Erosion', *Scientific American*, Offprint 817.

EMMETT, W. W. (1970). Op. cit. page 156 above.

ESU, F. (1966). 'Short Term Stability of Slopes in Unweathered Jointed Clays', *Géotechnique*, **16**, 321–8.

EYLES, R. J. and HO, R. (1970). 'Soil Creep on a Humid Tropical Slope', *Journal of Tropical Geography*, **31**, 40–2.

FLINT, J. J. (1973). 'Experimental Development and Headward Growth of Channel Networks', *Bulletin of the Geological Society of America*, **84**, 1087–94.

FLEMING, R. W. and JOHNSON, A. M. (1975). 'Rates of Seasonal Creep of a Silty Clay Soil', *Quarterly Journal of Engineering Geology*, **8**, 1–29.

GARDNER, J. (1969). 'Observations on Surficial Talus Movement', *Zeitschrift für Geomorphologie*, **13**, 317–23.

GARDNER, J. (1970). 'Rockfall—a Geomorphic Process in High Mountain Terrain', *Albertan Geographer*, **6**, 15–20.

GAUDIN, F. (1926) quoted in DREIMANIS, A. and VARGNERS, U. J. (1971). 'Bimodal Distribution of Rock and Mineral Fragments in Basal Tills', in *Till—a symposium*, ed. Goldthwait, Ohio State University Press.

GIBBS, R. J. (1967). 'The Geochemistry of the Amazon River System Part I: Factors Controlling the Salinity and the Composition and Concentration of the Suspended Solids', *Bulletin of the Geological Association of America*, **78**, 1203–32.

GIFFORD, R. O. and FRUGOLI, D. M. (1964). 'Silica Sources in Soil Solutions', *Science*, **145**, 386–8.

GOSS, D. W., SMITH, S. J. and STEWART, B. A. (1973). 'Movement of Added Clay through Calcareous Materials', *Geoderma*, **9**, 97–104.

HACK, J. T. and GOODLETT, J. C. (1960). 'Geomorphology and Forest Ecology of a Mountain Stream in the Central Appalachians', *United States Geological Survey Professional Paper 347*.

HALLSWORTH, E. G. (1963). 'An Examination of some Factors Affecting the Movement of Clay in an Artificial Soil', *Journal of Soil Science*, **14**, 360–71.

HANRAHAN, E. T. and PHILLIPS, M. (1976). 'Some Problems of Tunnelling in Dublin Boulder Clay', *Proceedings of the 6th European Conference on Soil Mechanics*, and *Foundation Engineering*, Vienna, in press.

HARVEY, D. (1969). Op. cit. page 31 above.

HATCH, F. H., WELLS, A. K. and WELLS, M. K. (1961). *Petrology of the Igneous Rocks*, Murby, London.

HENKEL, D. J. and SKEMPTON, A. W. (1955). 'A Landslide at Jackfield, Shropshire, in Heavily Overconsolidated Clay', *Géotechnique*, **5**, 131–7.

HJÜLSTROM, F. (1935). 'Studies of the Morphological Activities of Rivers as Illustrated by the River Fyris', *Bulletin of the Geological Institute, University of Uppsala*, **25**, 221–527.

HOLEMAN, J. N. (1968). 'The Sediment Yield of Major Rivers of the World', *Water Resources Research*, **4**, 737–47.

HOLMSEN, P. (1953). 'Landslides in Norwegian Quick Clays', *Géotechnique*, **3**, 187–94.

HOLTZ, W. G. (1960). 'The Effect of Gravel Particles on Friction', *Proceedings of the American Society of Civil Engineers—Research Conference on Shear Strength*, 1000–1.

HORTON, R. E. (1945). Op. cit. page 156 above.

HUTCHINSON, J. M. (1968). 'Field Meeting on the Coastal Landslides of Kent', *Proceedings of the Geologists' Association*, **79**, 227–37.

—— (1970). 'A Coastal Mudflow on the London Clay Cliffs at Beltinge, North Kent', *Geotechnique*, **20**, 412–38.

—— and BHANDARI, R. K. (1971). Op. cit. page 100 above.

JENNY, H. (1941). *Factors of Soil Formation*, McGraw-Hill, New York.

JOHNSON, A. M. and RAHN, P. H. (1970). 'Mobilisation of Debris Flows', *Zeitschrift für Geomorphologie*, Supplementband **9**, 168–86.

KENT, P. (1966). Op. cit. page 100 above.

KAWAKAMI, H. and ABE, H. (1970). 'Shear Characteristics of Saturated Gravelly Clays', *Transactions of the Japanese Society of Civil Engineers*, **2**, 295–8.

KERPEN, W. and SCHARPENSEEL, H. W. (1967). 'Movement of Ions and Colloids in Undisturbed Soil and Parent Rock Columns', *Proceedings of the International Atomic Energy Agency Symposium*, Istanbul, 213–55.

KIRKBY, A. V. T. and KIRKBY, M. J. (1974). Op. cit. page 156 above.

KIRKBY, M. J. (1967). Op. cit. page 113 above.

—— and CHORLEY, R. J. (1967). 'Throughflow Overland Flow and Erosion', *Bulletin of the International Association of Scientific Hydrologists*, **12**, 5–21.

KOJAN, E. (1967). 'Mechanics and Rates of Natural Soil Creep', *United States Forest Service Experimental Station Report*, 233–53.

KOONS, D. (1955). 'Cliff Retreat in the South West United States of America', *American Journal of Science*, **253**, 44–52.

KOVENYA, S. V., MELIKOVA, M. K. and FRID, A. S. (1972). 'A Study of the Role of Mechanical Forces and Geometric Conditions in the Movements of Highly Dispersed Particles in Soil Columns', *Soviet Soil Science*, **4**, 605–12.

LAMBE, T. W. and WHITMAN, R. V. (1969). *Soil Mechanics*, Wiley, New York.

LANGBEIN, W. B. and DAWDY, D. R. (1964). 'Occurrence of Dissolved Solids in Surface Waters in the United States', *United States Geological Survey Professional Paper 501-D*, 115–17.

—— and SCHUMM, S. A. (1958). 'Yield of Sediment in Relation to Mean Annual Precipitation', *Transactions of the American Geophysical Union*, **39**, 1076–84.

LAWS, J. O. and PARSONS, D. A. (1943). 'The Relation of Raindrop Size to Intensity', *Transactions of the American Geophysical Union*, **24**, 452–9.

LEOPOLD, L. B., WOLMAN, M. G. and MILLER, J. P. (1964). Op. cit. page 156 above.

LEWIS, L. A. (1974). 'Slow Movements of Earth under Tropical Rainforest Conditions', *Geology*, **2**, 9–10.

LOHNES, R. A. and HANDY, R. L. (1968). 'Slope Angles in Friable Loess', *Journal of Geology*, **76**, 247–58.

LOVERING, T. S. (1958). 'The Significance of Accumulator Plants in Rock Weathering', *Bulletin of the Geological Society of America*, **70**, 781–800.

MATHEWS, W. H. (1959). 'Vertical Distribution of Velocity in the Salmon Glacier, British Columbia', *Journal of Glaciology*, **3**, 448–67.

McCALL, J. G. (1960). 'The Flow Characteristics of a Cirque Glacier and Their Effects on Glacial Structure and Cirque Formation', in 'Norwegian Cirque Glaciers', *Royal Geographical Society Research Paper 4*, 34–62.

MILLER, J. P. (1961). 'Solutes in Small Streams Draining Single Rock-types, Sangre de Cristo Mountains, New Mexico', *United States Geological Survey Water Supply Paper*, 1535–F.

MOSELEY, H. (1869). 'On the Descent of a Solid Body on an Inclined Plane when Subjected to Alterations of Temperature', *Philosophical Magazine*, **38**, 99–118.

MOSELEY, M. P. (1973). Op. cit. page 156 above.

—— (1974). Op. cit. page 156 above.

NISHIDA, K. and KAGAWA, M. (1972). 'Shear Strength Properties of Weathered Residual Soils', *Technical Reports of Kansai University*, Osaka, **13**, 139–49.

NOBLE, H. L. (1973). 'Residual Strength and Landslides in Clay and Shale', *Journal of the Soil Mechanics and Foundation Engineering Division, Proceedings of the American Society of Civil Engineers*, **99**, 705–19.

PARRY, R. H. G. (1971). 'Stability Analyses for Low Embankments on Soft Clays', *Proceedings of the Roscoe Memorial Symposium, Stress-Strain Behaviour of Soils*, Cambridge University Press, London.

PATERSON, W. S. B. (1969). Op. cit. page 156 above.

PEEL, R. F. (1974). 'Insolation Weathering—Some Measurements of Diurnal Temperature Changes in Exposed Rocks in the Tibesti Region, Central Sahara', *Zeitschrift für Geomorphologie*, Supplementband **21**, 19–28.

PERUTZ, M. F. (1950). 'Direct Measurements of the Velocity Distribution in a Vertical Profile through a Glacier', *Journal of Glaciology*, **1**, 382–3.

PETTERSEN, S. (1958). *Introduction to Meteorology*, McGraw-Hill, New York.

PHIPPS, R. L. (1974). Op. cit. page 113 above.

POLYNOV, P. B. (1937). *The Cycle of Weathering*, Murby, London.

PRIOR, D. B. and STEPHENS, N. (1972). 'Some Movement Patterns of Temperate Mudflows: Examples from Northern Ireland', *Bulletin of the Geological Society of America*, **83**, 2533–44.

PRIOR, D. B., STEPHENS, N. and DOUGLAS, G. R. (1970). 'Some Examples of Modern Debris Flows from Northern Ireland', *Zeitschrift für Geomorphologie*, **14**, 275–88.

RAPP, A. (1960). 'Recent Developments of Mountain Slopes in Karkevagge and Surroundings, Northern Scandinavia', *Geografiska Annaler*, **42**, 65–200.

RENDON-HERRERO, O. (1974). 'Estimation of Wash Load Produced on Certain Small Watersheds', *Journal of the Hydraulics Division, Proceedings of the American Society of Civil Engineers*, **100**, 835–48.

RHOADES, R. and SINACORI, M. N. (1941). 'Patterns of Groundwater Flow and Solution', *Journal of Geology*, **49**, 785–94.

ROSENQVIST, T. (1953). 'Considerations on the Sensitivity of Norwegian Quick Clays', *Géotechnique*, **3**, 195–200.

ROWE, P. W. (1962). 'The Stress-Dilatancy Relation for Static Equilibrium for an Assemblage of Particles in Contact', *Proceedings of the Royal Society*, **A269**, 500–27.

RUDBERG, S. (1962). Op. cit. page 113 above.

RUHE, R. V. (1952). 'Topographic Discontinuities of the Des Moines Lobe', *American Journal of Science*, **250**, 216–56.

RUXTON, B. P. (1958). 'Weathering and Subsurface Erosion in Granite at the Piedmont Angle, Balos, Sudan', *Geological Magazine*, **95**, 353–77.

—— (1968). 'Measurements of the Degree of Chemical Weathering of Rocks', in *Landform Studies in Australia and New Guinea*, ed. J. N. Jennings and J. A. Mabbut, Cambridge University Press, London.

SCHEIDEGGER, A. E. (1961). 'Theory of Rock Movement on Scree Slopes', *Journal of the Alberta Society of Petroleum Geologists*, **9**, 131–40.

SCHUMM, S. A. (1956). Op. cit. page 156 above.

—— and CHORLEY, R. J. (1964). 'The Fall of Threatening Rock', *American Journal of Science*, **262**, 1041–54.

SEGERSTROM, K. (1950). 'Erosion Studies at Paracutin, State of Michoachan, Mexico', *United States Geological Survey Bulletin*, **965A.**

SEVALDSON, R. A. (1956). 'The Slide in Lodalen, October 6th 1954', *Géotechnique*, **6**, 167–82.

SHARP, R. P. (1942). 'Mudflow Levées', *Journal of Geomorphology*, **5**, 212–27.

SHARPE, C. F. S. (1938). *Landslides and Related Phenomena*, Columbia, New York.

SHREVE, R. L. (1966). 'Sherman Landslide, Alaska', *Science*, **154**, 1639–43.

—— (1968). 'Leakage and Fluidisation in Air-Layer Lubrication Avalanches', *Bulletin of the Geological Society of America*, **79**, 653–8.

SIMONETT, D. S. (1967). 'Landslide Distribution and Earthquakes in the Bewani and Torricelli Mountains, New Guinea: A Statistical Approach', in *Landform Studies in Australia and New Guinea*, ed. J. N. Jennings and J. A. Mabbut, Cambridge University Press, London.

SIMONS, D. B. (1956). 'A Study of the Angle of Repose of Non-Cohesive Materials', Unpublished Report, Department of Civil Engineering, Colorado State University.

SKEMPTON, A. W. (1948). 'The Rate of Softening in Stiff, Fissured Clays', *Proceedings of the 2nd International Conference on Soil Mechanics and Foundation Engineering*, **2**, 50–3.

—— (1964). Op. cit. page 100 above.

—— and BROWN, J. D. (1961). 'A Landslide in Boulder Clay at Selset, Yorkshire', *Géotechnique*, **11**, 280–93.

—— and DELORY, F. A. (1957). 'Stability of Natural Slopes in London Clay', *Proceedings of the 4th International Conference on Soil Mechanics and Foundation Engineering*, **2**, 378–81.

—— and NORTHEY, R. D. (1953). 'The Sensitivity of Clays', *Géotechnique*, **3**, 30–53.

SMALLEY, I. J. (1972). 'Boundary Conditions for Flowslides in Fine Particle Mine Waste Tips', *Transactions of the Institute of Mining and Metallurgy*, **81**, A31–A37.

SMITH, D. I. (1965). 'Some Aspects of Limestone Solution in the Bristol Region', *Geographical Journal*, **131**, 44–9.

STATHAM, I. (1974). 'The Relationship of Porosity and Angle of Repose to Mixture Proportions in Assemblages of Different Sized Materials', *Sedimentology*, **21**, 149–62.

—— (1975). 'Slope Instabilities and Recent Slope Development in Glencullen, County Wicklow. *Irish Geography*, **8**, 42–54.

—— (1976). 'Debris Flows on Vegetated Screes in the Black Mountain, Carmarthenshire', *Earth Surface Processes*, **1**, 173–80.

SVASEK, J. J. and TERWINDT, J. H. J. (1974). 'Measurement of Sand Transport on a Natural Beach', *Sedimentology*, **21**, 739–42.

TAMBURI, A. J. (1974). Op. cit. page 114 above.

TER-STEPHANIAN, G. (1965). 'In-situ Determination of the Rheological Characteristics of Soils on Slopes', *Proceedings of the 6th International Conference on Soil Mechanics and Foundation Engineering*, 575–7.

TERZAGHI, K. (1950). 'The Mechanism of Landslides', *Bulletin of the Geological Society of America, Berkeley Volume*, 83–122.

—— (1962). Op. cit. page 100 above.

UNESCO: *Annual Summaries of Natural Disasters*, UNESCO, Paris.

VAUGHAN, P. R. and WALBANCKE, H. J. (1973). 'Pore-Pressure Changes and the Delayed Failure of Cutting Slopes in Overconsolidated Clay', *Géotechnique*, **23**, 531–41.

WARREN, A. (1974). 'Observations on Dunes and Bimodal Sands in the Ténére Desert', *Sedimentology*, **19**, 37–44.

WENTWORTH, C. K. (1943). Op. cit. page 100 above.

WHALLEY, W. B. (1974). 'The Mechanics of High Magnitude, Low Frequency Rock-Failure and its Importance in Mountainous Areas', *University of Reading, Geography Papers*, **27**, 48pp.

WILLIAMS, M. A. J. (1968). 'Termites and Soil Development near Brocks Creek, Northern Territories', *Australian Journal of Science*, **31**, 153–4.

—— (1969). 'Prediction of Rainsplash in the Seasonally Wet Tropics', *Nature*, **222**, 763–5.

—— (1974). 'Surface Rock-Creep on Sandstone Slopes in Northern Territory, Australia', *Australian Geographer*, **12**, 419–24.

WILLIAMS, P. J. (1966). 'Downslope Movement at a Subarctic Location with Regard to its Variation with Depth', *Canadian Geotechnical Journal*, **3**, 191–203.

WILSON, I. G. (1971). Op. cit. page 156 above.

WINDER, C. G. (1965). 'Alluvial Cone Construction by an Alpine Mudflow in a Humid Temperate Region', *Canadian Journal of Earth Sciences*, **2**, 270–7.

WOLMAN, M. G. and MILLER, J. P. (1960). Op. cit. page 176 above.

YEN, B. C. (1969). Op. cit. page 114 above.

YOUNG, A. (1960). 'Soil Movement by Denudational Processes on Slopes', *Nature*, **188**, 120–2.

—— (1974). Op. cit. page 114 above.

Index